European Guide to Power System Testing

Thomas I. Strasser · Erik C. W. de Jong ·
Maria Sosnina

Editors

European Guide to Power System Testing

The ERIGrid Holistic Approach
for Evaluating Complex Smart Grid
Configurations

OPEN

 Springer

Editors
Thomas I. Strasser
Center for Energy
AIT Austrian Institute of Technology
Vienna, Austria

Erik C. W. de Jong
KEMA Labs – Flex Power Grid Lab
KEMA B.V.
Arnhem, The Netherlands

Maria Sosnina
European Distributed Energy Resources
Laboratories (DERlab) e. V.
Kassel, Germany

ISBN 978-3-030-42273-8   ISBN 978-3-030-42274-5   (eBook)
https://doi.org/10.1007/978-3-030-42274-5

This Springer imprint is published by the registered company Springer Nature Switzerland AG
The registered company address is: Gewerbestrasse 11, 6330 Cham, Switzerland

Preface

Future Power Systems—A Complex System of Systems

A driving force for the realization of a sustainable energy supply in Europe is the integration of distributed, renewable energy resources. Due to their dynamic and stochastic generation behaviour, utilities and network operators are confronted with a more complex operation of the underlying power grids. Additionally, due to the higher flexibility on the consumer side through partly controllable loads, ongoing changes of regulatory rules, technology developments and the liberalization of energy markets, the system's operation needs adaptation. Sophisticated design approaches together with proper operational concepts and intelligent automation provide the basis to turn the existing power system into a cyber-physical energy system, a so-called "Smart Grid".

Whereas transmission systems are already well equipped with sophisticated measurement devices and are centrally operated, the integration of renewable generators occurs mainly at the level of distribution networks, which used to be operated in a passive way until now. With smaller units at this grid level, the number of sub-systems and devices to be monitored and controlled is steadily increasing. While the application of modern scalable information and communication technology facilitates this integration, it also creates further coupling of engineering domains where little mutual interdependencies existed before. Challenged by this development, new methodologies and practices must be developed. Viewing the electric energy infrastructure in its entirety as a cyber-physical, critical infrastructure, such new methodologies and practices will have to ensure that the classical high-reliability, real-time operation, and regulatory requirements can be met also in the future.

Who Should Read This Book

The observed increase of complexity thus manifests in increased coupling across domains, such as electricity, automation and information technology, but also in scale and heterogeneity. The effects of this are not trivial to anticipate, nor to summarize: operational aspects become a concern, as failures may propagate across increasingly interdependent automation systems; energy management and coordination may become more challenging. Another challenge with complexity is the ability to conceive, design and develop critical infrastructure systems that depend on such cross-disciplinary competences.

Before deployment in an operational environment, smart grid solutions and products have to be validated and tested. Industry and researchers have recognized this challenge, and a growing number of smart grid projects around the world have led to a significant portfolio of developments and demonstrations. However, the status quo for power systems testing (which is an integral part of the development process) is to focus mainly on the device-level, meanwhile simplifying the behaviour of other components to electrical equivalents. This traditional decoupling raises a question of the global behaviours of the integrated system. A combination of different technologies across domains requires that communication among specialists is established and founded on the interconnection of different disciplines during the development process. The heterogeneity of smart grid domains and technologies, notably the interactions between the various technologies, conflicts with the traditional approach: test labs often specialize in a certain domain and can hence only test components for a particular sub-system. In order to support the different stages of the overall development process for smart grid solutions, tests are needed to evaluate the integration on a system level, addressing all relevant test domains. Proposed alternative testing approaches include virtual (simulation) or semi-virtual (hardware-in-the-loop) experiments that cover multiple domains. For these new approaches, questions arise as to whether the test results can be considered valid enough to draw firm conclusions for a real-world deployment of the tested systems.

Since the validation of smart grid solution on the system level is not common until now and corresponding approaches, concepts and tools are currently in development, this book provides an overview of the achievements and results which have been obtained in the European ERIGrid project. This book targets professionals and engineers but also researchers and young students active in the domain of power and energy systems dealing with the development and validation of new applications, solutions and technologies.

Contribution

This book summarizes the main achievements and results from the European research infrastructure project ERIGrid (supported by the European Commission under Grant Agreement No. 654113) related to power system/smart grid validation and testing on the system level which has been carried out during the last 4.5 years (i.e. November 2015 to April 2020) by 18 partners distributed in 11 European countries.

In the following chapters, the developed validation approaches, simulation, hardware-in-the-loop and laboratory-based testing concepts (including coupling of research infrastructures/laboratories) are being discussed in detail. Moreover, their application on selected scenarios and test cases are being demonstrated. Furthermore, lessons learned from the usage of the aforementioned tools are being provided. Besides the validation approaches and tools also concepts for the education and training on smart grid topics are introduced. Finally, the book is concluded with the summary of the achievements as well as with an outlook about necessary future research and development.

The project results are usually available online at the corresponding project website https://erigrid.eu and the website of the funding agency https://cordis. europa.eu/project/id/654113. The content of this book is based on it, and most parts are drawn from the relevant project deliverables and publications.

Vienna, Austria
Arnhem, The Netherlands
Kassel, Germany
April 2020

Thomas I. Strasser
Erik C. W. de Jong
Maria Sosnina

Acknowledgements We are very grateful to all the ERIGrid partners which contributed to this book, especially to the work package and task leaders D. Babazadeh, P. Teimourzadeh Baboli (OFFIS), K. Heussen (DTU), Arjen A. van der Meer (TU Delft), Van Hoa Nguyen (CEA), L. Pellegrino (RSE), K. Maki (VTT) and P. Kotsampopoulos (NTUA) who coordinated the writing of the corresponding book chapters. Special thanks go also to E. Mrakotsky-Kolm (AIT) for reviewing the whole book. Moreover, we want to provide our thanks to K. R. Selvaraj, A. Doyle and W. Hermens from Springer for their great help and support during the writing and editing of this work. Finally, we want to thank the European Commission for providing the financial support for this work which made the corresponding results possible.

Contents

Acronyms

AC	Alternating Current
CHIL	Controller Hardware-in-the-Loop
CHP	Combined Heat and Power
CIM	Common Information Model
CPES	Cyber-Physical Energy System
CVC	Coordinated Voltage Control
DC	Direct Current
DoE	Design of Experiment
DRTS	Digital Real-Time Simulator
DSO	Distribution System Operator
DuI	Domains under Investigation
DUT	Device under Test
ES	Experiment Specification
FMI	Functional Mock-up Interface
FMU	Functional Mock-up Unit
GUI	Graphical User Interface
HIL	Hardware-in-the-Loop
HTD	Holistic Test Description
HUT	Hardware-under-test
HV	High Voltage
HVDC	High Voltage Direct Current
ICT	Information and Communication Technology
LSS	Large Scale System
LV	Low Voltage
MV	Medium Voltage
NIIPS	Non-Interconnected Islanded Power System
OLTC	On-load Tap Changing Transformer
OuI	Object under Investigation
PHIL	Power Hardware-in-the-Loop
PMU	Phasor Measurement Unit

PoI	Purpose of Investigation
PSIL	Power System-in-the-Loop
PV	Photovoltaic
QS	Qualification Strategy
R&D	Research and Development
RES	Renewable Energy Source
RI	Research Infrastructure
SCADA	Supervisory Control and Data Acquisition
SGAM	Smart Grid Architecture Model
SIL	Software-in-the-Loop
SuT	System under Test
TC	Test Case
TS	Test Specification
TSO	Transmission System Operator
UFLS	Under Frequency Load Shedding
VRI	Virtual Research Infrastructure

Towards System-Level Validation

T. I. Strasser⊚, F. Pröstl Andrén, M. Calin, E. C. W. de Jong, and M. Sosnina

1 Higher Complexity in Future Power Systems

Power system operation is of vital importance and has to be developed far beyond today's practice in order to meet future needs like the integration of renewables or battery storage systems [3]. In fact, nearly all European countries faced an abrupt and very important growth of Renewable Energy Sources (RES) such as wind and photovoltaic that are intrinsically variable and up to some extent difficult to predict. In addition, an increase of new types of electric loads such as air conditioning, heat pumps, and electric vehicles; and a reduction of traditional generation power plants can be observed. Hence, the level of complexity of system operation increases steadily. To avoid dramatic consequences, there is an urgent need for a system flexibility increase [18]. Also the roll-out of smart grids applications and solutions such as Information and Communication Technology (ICT) and power electronic-based grid components is of particular importance in order to realize a number of advanced system functionalities (power/energy management, demand side management, ancillary services provision, etc.) [6, 7, 13, 14].

In such a Cyber-Physical Energy System (CPES)—also denoted as "Smart Grid" in the literature [3]—this also requires distributed intelligence on different levels in the system as outlined in Fig. 1 and Table 1. Flexibility, adaptability, scalability, and autonomy are key points to realize the automation systems and component controllers of CPES [13]. Also, interoperability and open interfaces are important to enable the above described functions on the different levels of intelligence [6]. Hence, such kind

T. I. Strasser (✉) · F. Pröstl Andrén · M. Calin
AIT Austrian Institute of Technology, Vienna, Austria
e-mail: thomas.strasser@ait.ac.at

E. C. W. de Jong
KEMA B.V., Arnheim, The Netherlands

M. Sosnina
European Distributed Energy Resources Laboratories (DERlab) e.V., Kassel, Germany

© The Author(s) 2020
T. I. Strasser et al. (eds.), *European Guide to Power System Testing*,
https://doi.org/10.1007/978-3-030-42274-5_1

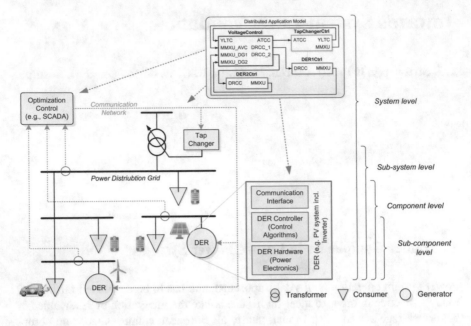

Fig. 1 Intelligence on different levels of smart grid systems [13, 14]

of systems tend to have a much higher complexity compared to traditional power systems [14].

2 Needs for System-Level Validation

2.1 Engineering and Validation Process

Typically, the engineering and validation process of CPES applications and solutions involves several stages depending on the applied design methodology or process model (V-model, etc.). Also, the overall complexity of the system under development usually influences the whole process (e.g., the development of a micro-grid controller is less complex as the supervisory control of a power distribution grid). However, the four main stages are common for the whole process as depicted in Fig. 2 and described in Table 2 [14, 17]. As indicated in the figure a step back to an earlier stage is possible if necessary. This can happen if the requirements are not met in a certain stage and a refinement of the previous one is necessary.

Table 1 Smart grid systems and their elements/components (adopted from [13, 14])

Level	Type of intelligence
System	Power system operation applications like energy management, distribution management or demand-side management are tackled at this level. Services of the underlying sub-systems and components are executed in a coordinated manner
Sub-systems	Control and optimization are the main tasks whereas the corresponding functions and algorithms have to deal with a limited number of components (generators, storages, etc.). Micro-grid control approaches as well as building energy management concepts are typical examples for this level
Components	Distributed Energy Resources (DER)/RES, battery storage systems or electric vehicle supply equipment is covered by this level. Such devices typically provide advanced functions like ancillary services (reactive power and voltage control, inertia and frequency control, etc.). Intelligence is either used for local optimization purposes (device behaviour) or for the optimization of systems/sub-systems on higher levels in a coordinated manner
Sub-components	Intelligence is used to improve the local component behaviour (harmonics, flicker, etc.). Power electronics and their advanced control functions is the driver for local intelligence. The controllers of DER, energy storages and other type of power system equipment (tap-changing transformers, smart breakers, etc.) can be considered as examples for sub-components

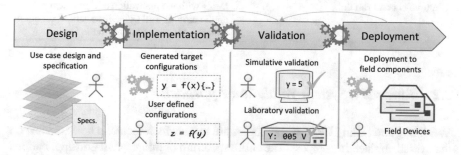

Fig. 2 Overview of the design and validation process of CPES applications [10]

Compared to other domains, the main challenges during the engineering and validation process of CPES applications and solutions can be summarized as [14]: *(i)* the fulfilment of high-reliability requirements, *(ii)* the observance of (strict) real-time requirements, *(iii)* the compliance with national rules, and *(iv)* the interaction with several system integrators/manufacturers. In order to prove the outcomes and results of the different stages proper concepts, methods, and corresponding tools are required. Due to the higher complexity of smart grid systems advanced testing methods are necessary addressing cyber-physical and multi-domain issues.

Table 2 Main design and validation stages for CPES development (adopted from [14, 17])

Process stage	Description of activities
System-level requirements and design	The system-level requirements and application scenarios are being identified (i.e., use cases). In the following a basic design and high-level architecture specification are typically carried out. After the conceptual design has been elaborated a detailed engineering of the system under development is done. Functions and services are also identified and specified
Implementation and prototype development	Usually prototypes are being developed at this stage. The process of transforming a concept into a prototype often introduces issues which were not considered during the design stage(s). Often boundary problems like communication latencies or non-linearities are neglected during the first versions of a basic concept. During the development of a prototype iterative refinements of solutions/algorithm are often necessary
System validation and component testing	After the first prototypes are available they are being tested (often either in simulation or in a laboratory environment). Test are usually carried out at component level first and afterwards integration tests are being performed
Deployment and roll out	Deals with the realization of a product or application as well as the installation/roll out of components and solutions in the field

2.2 Towards a System Validation Approach

Validating and testing CPES technologies are tasks which require a holistic view on the overall engineering process. The entire spectrum of future smart grid applications and solutions has to be taken into consideration, but also the whole engineering process (as depicted in Fig. 2). Even more, the whole range of aspects from interest and relevance for a stable, safe and efficient smart grid system has to be regarded [2, 14]. Comparable processes have already been successfully implemented in other application domains like automotive, consumer electronics, mechanical/chemical engineering, albeit on an arguably less complex level [1]. The domain of power and energy system can profit from existing approaches and can adapt them to fulfil needs and requirements of the domain. There is no need to start from scratch.

However, until now there is a lack of an integrated approach for the engineering and validating CPES covering power system, ICT as well as automation and control aspects in an integrated manner. Several mandatory testing approaches are nowadays available, but they are mainly focused on the device-level [2]. Those approaches are usually not sufficient to test a whole CPES configuration on the system level [14]. In order to guarantee a sustainable and secure supply of electricity in a more complex smart grid system as well as to support the expected forthcoming large-scale roll out of new technologies, proper validation and testing methods are necessary. They need to cover the power system in a cyber-physical and multi-domain manner. Therefore, the following needs can be identified [10, 14]:

- *Cyber-physical, multi-domain approach:* System integration topics including analysis and evaluation need to be addressed on the system level in a cyber-physical and multi-domain manner.
- *Holistic validation framework:* A suitable framework which allows the holistic analysis and evaluation of CPES approaches on the system level is required; this also includes the corresponding Research Infrastructure (RI).
- *Standardized procedures:* Harmonized and possibly standardized validation procedures and tools need to be developed.
- *Educated professionals:* Besides the technical validation aspects, engineers and researchers need to be properly educated in order to understand smart grid solutions in a cyber-physical and multi-domain manner. They need to be aware about the main testing requirements.

2.3 Illustrative Example

For a better understanding of future system validation needs a coordinated voltage control in a power distribution grid is introduced [14]. Figure 1 provides an overview of this illustrative smart grid example where an On-Load Tap Changing (OLTC) transformer is used together with reactive and active power control provided by DERs (e.g., photovoltaic generator and small wind turbines) and battery storage systems. The goal of this application is to keep the voltage in the grid in defined boundaries and therefore to increase the hosting capacity of renewables [12]. The corresponding control approach has to calculate the optimal position of the OLTC and to derive set-points for reactive and active power of DER units. Those control commands are usually send over a communication network to the corresponding components.

Before deploying this solution into the field various tests need to be carried out like the validation of the different components (incl. local control approaches and communication interfaces) on the sub-component and component level. Also, the local OLTC control approach needs to be evaluated, too. Nevertheless, the integration of all components and sub-systems is one of the most important issues. The proper functionality of all components is not a guarantee that the whole system is behaving as expected. As outlined above, a system-level testing is required in order to prove that the whole application (addressing power system and ICT topics) is working properly and as expected [14].

This example will be used also later in the book for the explanation of the developed validation methods and corresponding testing tools.

Table 3 Overview of validation and testing approaches (adopted from [14])

Method	Stage			
	Basic design	Detailed design	Prototype	Deployment
Analytics and simulation	+	++	o	−
Real-time sim. and HIL	−	−	++	+
Lab-based testing	−	−	++	++
Field trials	−	−	−	++

Legend: − …less suitable, o …suitable with limitations, + …suitable, ++ …best choice

3 Existing Approaches and Research Directions

3.1 Suitable Methods and Tools

In the literature there are a several well-known development and validation methods documented which are suitable for the domain of power and energy systems [14, 17]. The most promising approaches are *(i)* analytic analysis and software simulation, *(ii)* real-time simulation and Hardware-in-the-Loop (HIL), *(iii)* laboratory-based testing, and *(iv)* field trials and large-scale demonstration projects. However, they are useful for specific activities and they are not covering all process stages as outlined in Fig. 2 in the same way. Therefore, Table 3 provides an overview where those methods fit best and where not.

Simulation-based approaches are very common in power systems engineering. Individual technological areas (power system, ICT/automation) have been analysed in dedicated simulation tools. Transient stability and steady state simulations are very often used to investigate the behaviour of power systems and their components where various tools are nowadays available [8]. Comparable developments can be observed also in the domain of ICT and automation systems.

Nevertheless, the development of CPES applications urge for a more integrated simulation approach covering all targeted areas. The usage of simulation as development approach gets more of interest. Analysing the behaviour of smart grid systems requires hybrid models combining continuous time-based (physics-related) and discrete event-based (communication and controls-related) aspects. Co-simulation (or co-operative simulation) is an approach for the joint simulation of models developed with different tools (tool coupling) where each tool treats one part of a modular coupled problem. Co-simulation takes under consideration the complexity of the simulated system and influences between different aspects or domains interconnected in the same system [9, 11, 14].

Nowadays HIL-based approaches get more interest from the power system domain. Two different approaches can be distinguished, namely Controller-Hardware-in-the-Loop (CHIL) and Power-Hardware-in-the-Loop (PHIL). The first approach is used to evaluate a controller platform and the corresponding algorithm(s) where the real implementation is available and the power system component is simulated in a real-time environment. Besides that, PHIL provides a more advanced tool for power system analysis, testing and validation by evaluating the actual power device with a real-life system which is simulated in a Digital Real-Time Simulator (DRTS), allowing repeatable and economical testing under realistic, highly flexible and scalable conditions. Extreme conditions can be studied with minimum cost and risk, while problematic issues in the equipment behaviour can be revealed allowing an in depth understanding of the tested device. PHIL testing combines the benefits of numeric simulation and hardware testing and is constantly gaining interest at international level [4, 5, 14].

Laboratory experiments in electrical engineering for testing or certifying single or small setups of components are common practice. However, the decentralization of operation and control as well as the massive deployment of ICT components (and thus introducing shorter innovation and product cycles than hitherto known in energy supply systems) drastically increase the complexity of the system under investigation and easily exceed the scope of existing laboratory setups. In a single laboratory environment, the evaluation of holistic CPES is out of the question leaving simulation (or hybrid co-simulation experimentation incl. HIL) setups as the only viable option. Moreover, flexibility to deploy intelligent algorithms in different locations across the system is also necessary to move towards a laboratory-based testing of integrated power systems. Another issue is that laboratories are often developed for a specific purpose and they cannot be adapted easily (from the technical but also from the financial point of view) [2, 14].

Besides simulation and lab-based validation approaches (incl. HIL) field trials and large-scale demonstration projects are also of importance for the validation of new architectures and concepts. They have the advantage to test industrial-like prototypes and developments under real-world conditions, but a huge amount of preparation and planning work is necessary to realize such kind of field trials. Usually, they are also quite expensive and resource intensive [14].

3.2 Future Research Directions

As outlined above there are a couple of interesting approaches available in the literature which are suitable for validation and testing. However, all of them have in common that they are usually address a specific domain and they are not really developed to cope with the cyber-physical and multi-domain nature of CPES applications and solutions. In order to analyse and evaluate such a multi-domain configuration, a set of corresponding methods, procedures, and corresponding tools are necessary. Usually, pure virtual-based methods are not enough for validating smart grid systems, since the availability of proper and accurate simulation models cannot always be guaranteed (e.g., inverter-based components are some-times very complex to model or it takes too long to get a proper model). Simulation and lab-based validation approaches have to be combined and used in an integrated manner covering the

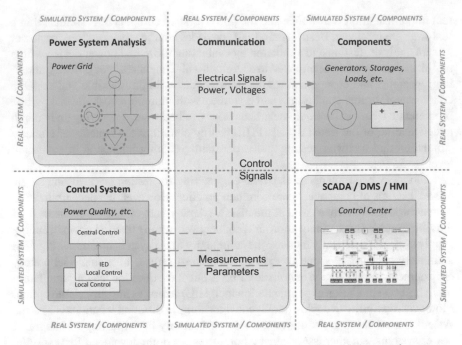

Fig. 3 Integrated CPES validation using virtual and real components (adopted from [14, 16])

whole range of opportunities and challenges. Such an approach is necessary when answering system level integration and validation questions [14].

Figure 3 sketches this idea where a flexible combination of physical components (available in a laboratory environment) and simulation models are combined in a flexible way in dependence of the corresponding validation or testing goal. Such an approach needs the improvement of available methods and tools. In addition, proper interfaces need to be provided as well. There is still space for future research and development related to this topic [14].

Besides that, system-level validation procedures as well as corresponding benchmark criteria are necessary. Moreover, also the linking of existing RIs as well as the establishment of clusters of them should be in the focus of future research. Such an integrated RI should be able to provide advanced validation and testing services fulfilling future validation needs in a cyber-physical manner. Last but not least also the training and education of engineers and researchers active in the domain of power and energy systems need to be educated on CPES topics [14].

4 Overview of the ERIGrid Validation Approach

To overcome the shortcomings in power system evaluation as briefly outlined above the ERIGrid approach has been developed. It addresses the open points by developing a holistic, cyber-physical systems-oriented approach for testing smart grid systems. This integrated European smart grid RI targets the following points [15]:

- Creation of a single point of reference promoting research, technology development, and innovation on all aspects of smart grid systems validation,
- Development of a coordinated and integrated approach using the partners' expertise and infrastructures more effectively, adding value to research projects, and promoting European leadership in smart grid systems,
- Facilitating a wider sharing of knowledge, tools, and techniques across fields and between academia and industry across Europe, and
- Accelerating pre-normative research and promoting the rapid transfer of research results into industrial-related standards to support future smart grids development, validation and roll out.

To realize the above introduced project goals the following main research and development activities have been identified for the ERIGrid project:

- Development of a formalized, holistic validation procedure for testing smart grid systems and corresponding configurations,
- Improvement of simulation and lab-based testing methods supporting the validation activities, and
- The provision of a corresponding and integrated pan-European RI based on the partner's laboratories.

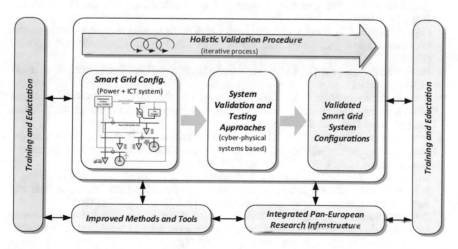

Fig. 4 Overview of the ERIGrid approach [15]

Additionally, training and education concepts are also being developed to support the overall research activities. An interesting point in the ERIGrid approach is to provide free access to the integrated RI (i.e., partner's smart grid laboratories) and the corresponding methods and tools for external user groups from industry and academia as outlined in Fig. 4.

The holistic testing methodology should facilitate conducting tests and experiments representative of integrated smart grid systems by testing and experimentation across distributed RIs, which might not necessarily be functionally interconnected.

In the following chapters main validation concepts and procedures as well as the corresponding tools are introduced and demonstrated on selected scenarios. Lessons learned and educational approaches are discussed as well.

References

1. Bringmann, E., Krämer, A.: Model-based testing of automotive systems. In: 2008 1st International Conference on Software Testing, Verification, and Validation, pp. 485–493. IEEE (2008)
2. Bründlinger, R., Strasser, T., Lauss, G., Hoke, A., et al.: Lab tests: verifying that smart grid power converters are truly smart. IEEE Power Energy Mag. 13(2), 30–42 (2015)
3. Farhangi, H.: The path of the smart grid. IEEE Power Energy Mag. 8(1), 18–28 (2010)
4. Faruque, M.D.O., Strasser, T., Lauss, G., Jalili-Marandi, V., et al.: Real-time simulation technologies for power systems design, testing, and analysis. IEEE Power Energy Technol. Syst. J. 2(2), 63–73 (2015)
5. Guillaud, X., Faruque, M.O., Teninge, A., Hariri, A.H., et al.: Applications of real-time simulation technologies in power and energy systems. IEEE Power Energy Technol. Syst. J. 2(3), 103–115 (2015)
6. Gungor, V.C., Sahin, D., Kocak, T., Ergut, S., Buccella, C., Cecati, C., Hancke, G.P.: Smart grid technologies: Communication technologies and standards. IEEE Trans. Indust. Inform. 7(4), 529–539 (2011)
7. Liserre, M., Sauter, T., Hung, J.Y.: Future energy systems: integrating renewable energy sources into the smart power grid through industrial electronics. IEEE Indust. Electron Mag. 4(1), 18–37 (2010)
8. Milano, F.: An open source power system analysis toolbox. IEEE Trans. Power Syst. 20(3), 1199–1206 (2005)
9. Palensky, P., van der Meer, A.A., López, C.D., Jozeph, A., Pan, K.: Co-simulation of intelligent power systems - fundamentals, software architecture, numerics, and coupling. IEEE Indust. Electron. Mag. 11(1) (2017)
10. Pröstl Andrén, F., Strasser, T., Seitl, C., Resch, J., Brandauer, C., Panholzer, G.: On fostering smart grid development and validation with a model-based engineering and support framework. In: Proceedings of the CIRED Workshop 2018 (2018)
11. Steinbrink, C., Lehnhoff, S., Rohjans, S., Strasser, T.I., et al.: Simulation-based validation of smart grids–status quo and future research trends. In: International Conference on Industrial Applications of Holonic and Multi-Agent Systems, pp. 171–185. Springer (2017)
12. Stifter, M., Bletterie, B., Brunner, H., Burnier, D., et al.: Dg demonet validation: voltage control from simulation to field test. In: 2011 2nd IEEE PES International Conference and Exhibition on Innovative Smart Grid Technologies, pp. 1–8 (2011)
13. Strasser, T., Andrén, F., Kathan, J., Cecati, C., et al.: A review of architectures and concepts for intelligence in future electric energy systems. IEEE Trans. Indust. Electron. 62(4), 2424–2438 (2015)

14. Strasser, T., Pröstl Andrén, F., Lauss, G., Bründlinger, R., et al.: Towards holistic power distribution system validation and testing – an overview and discussion of different possibilities. e & i Elektrotechnik und Informationstechnik **134**(1) (2017)
15. Strasser, T., Pröstl Andrén, F., Widl, E., Lauss, G., et al.: An integrated pan-European research infrastructure for validating smart grid systems. e & i Elektrotechnik und Informationstechnik **135**(8), 616–622 (2018)
16. Strasser, T., Stifter, M., Andrén, F., Palensky, P.: Co-simulation training platform for smart grids. IEEE Trans. Power Syst. **29**(4), 1989–1997 (2014)
17. Strasser, T.I., Andrén Pröstl, F.: Engineering and validating cyber-physical energy systems: needs, status quo, and research trends. In: International Conference on Industrial Applications of Holonic and Multi-Agent Systems, pp. 13–26. Springer (2019)
18. Villar, J., Bessa, R., Matos, M.: Flexibility products and markets: literature review. Electric Power Syst. Res. **154**, 329–340 (2018)

Test Procedure and Description for System Testing

K. Heussen, D. Babazadeh, M. Z. Degefa, H. Taxt, J. Merino, V. H. Nguyen, P. Teimourzadeh Baboli, A. Moghim Khavari, E. Rikos, L. Pellegrino, Q. T. Tran, T. V. Jensen, P. Kotsampopoulos, and T. I. Strasser ⓘ

1 Introduction

System-level validation of smart grid solutions can be a complex effort. A typical smart grid solution, such as a distribution grid centralized demand response control system encompasses multiple disciplines (market, ICT, automation, infrastructure) and physical infrastructures (e.g. electricity, communication networks). Interactions

K. Heussen (✉) · T. V. Jensen
Technical University of Denmark, Roskilde, Denmark
e-mail: kh@elektro.dtu.dk

D. Babazadeh · P. Teimourzadeh Baboli
OFFIS – Institute for Information Technology, Oldenburg, Germany

M. Z. Degefa · H. Taxt
SINTEF Energi AS, Trondheim, Norway

J. Merino
TECNALIA Research & Innovation, Derio, Spain

V. H. Nguyen · Q. T. Tran
Université Grenoble Alpes, INES, Le Bourget du Lac, France and CEA, LITEN, Le Bourget du Lac, France

A. Moghim Khavari
DERlab, Kassel, Germany

E. Rikos
Centre for Renewable Energy Sources and Saving, Athens, Greece

L. Pellegrino
Ricerca Sistema Energetico, Milan, Italy

P. Kotsampopoulos
National Technical University of Athens, Athens, Greece

T. I. Strasser
AIT Austrian Institute of Technology, Vienna, Austria

© The Author(s) 2020
T. I. Strasser et al. (eds.), *European Guide to Power System Testing*,
https://doi.org/10.1007/978-3-030-42274-5_2

13

among automation systems, enabling ICT, and electricity infrastructure are in the nature of such solutions and make testing the integrated system a necessity.

As motivated in Sect. 1.2.2, appropriate testing for such Cyber-physical Energy Systems (CPES) is challenging as it requires availability of multi-disciplinary engineering expertise, as well as suitable tool integration regarding the testing platforms [17]. A re-organization of testing practices in research and industry is ongoing to harvest the benefits of the advanced integration of system components using suitable testing tool chains and workflows.

In this chapter, we aim to support this re-organisation of testing practice, by offering answers to the following questions:

i. How can system validation efforts be framed as experiments in order to account for complex system requirements and functions, the multi-disciplinary experts, and the wide variety of employed experimental platforms?

ii. What information is necessary to record in an experiment description, to fully document purpose, structure and execution of experiments for coordinated both planning and reporting purposes?

This chapter offers a viewpoint for harmonization of system validation efforts by focusing in the problem of test formulation. Considering question *i.*, At first, the problem of system testing is formulated, which leads to a generalized procedural pattern, to be called 'holistic testing procedure', introduced in Sect. 2. Here 'holistic' refers to the procedure's generality, as it should be, in principle, applicable to simple as well as very complex testing problems. To address question *ii.*, a test description method is introduced in Sect. 3 which is based on the named procedure.

1.1 Testing Procedure and Test Description

In the smart energy domain, a significant attention has been given to the abstract and structured description of system solution requirements, e.g. with use cases and SGAM [1]. However, abstract requirements specification is insufficient to derive test descriptions immediately. A "test specification gap" can be identified between those requirements and the structured preparation of validation efforts. And this gap further increases with increasing complexity of cyber-physical system structure of solutions, as well as advancements in test platform technology.

1.2 Holistic Testing for System Validation

A clear and formalized test description can improve the reusability and reproducibility of tests. It can facilitate both the preparation and execution of tests in spite of increasing complexity due to multi-domain systems and advanced experimental platforms. A structured approach also helps the identification of relevant test parameters and targets involving multiple domains. A speedup is also needed in R&D activities that require component characterization and system validation experiments.

To frame the problem of dealing with workflows and tool chain integration for testing, we define:

> **Holistic testing** is the process and methodology for the evaluation of a concrete function, system or component (object under investigation) within its relevant operational context (system under test), as required by the test objectives.

Here, *Object under Investigation (OuI)* is the component (hardware or software) that is subject to the test objective(s). Note that in system validation, there can be a number $n \geq 2$ of OuIs. The concept of OuI replaces related concepts used in practice, such as "device under test" (commonly abbreviated DUT), or "equipment under test".

The *System under Test (SuT)* refers to the system configuration that includes all relevant behaviors and interactions that are required to examine the test objectives. The OuI is thus a subset of the SuT, and the remaining aspects of the SuT are simulated, emulated, or realised by the testing platform.

The holistic testing concept thus provides a scaffold for the formulation of procedures, description methods and tool chains for testing:

- *procedures* take a user in steps through a testing campaign, sequencing tasks and outcomes appropriately;
- *description methods* ask the right questions and structure the outcomes in a harmonized and with a common interpretation;
- *tool chains* support and integrate the workflows and descriptions with suitable test platforms.

The approach to test description presented here is based on three basic aspects of testing: *(i)* The object and purpose of test (i.e. What is tested and why), *(ii)* the test elements and test protocol, and *(iii)* the physical or virtual facility (i.e. test platform) employed to realize the experiment.

In this vision, the scoping and design of validation tests and experiments is facilitated by offering a better formal framing and a procedural guideline.

2 Toward Procedures for System Validation

The need for system validation has been previously expressed, and holistic testing has been formulated as a concept to organise procedures, tools and descriptions. In this section the procedural view on the system validation problem is introduced by first discussing the role of testing in the development context, introducing a 'holistic' procedural view on testing, and finally presenting a specific procedure for integrating development and testing with different test platforms.

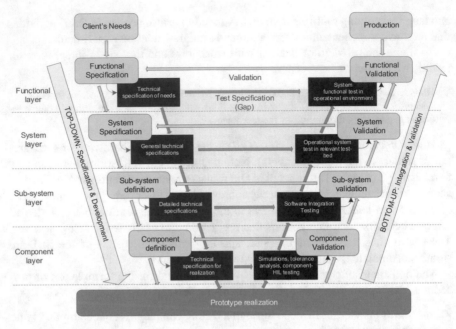

Fig. 1 Specification and testing layers in the conventional V-model, and the stipulated "test specification gap"

2.1 Purpose of Testing in the Development Process

Experiments play a role in the early stages of a technical design as well as in the final stages where technical solutions are evaluated against technical specifications and system level requirements. Systems design processes in industry follow the general scheme of the V-model [9], as mentioned in Sect. 2.1. This V-model can be interpreted classically as *waterfall* sequential process, but can also be used for modern concurrent engineering as a conceptual hierarchy, where the V-model establishes a strong coupling of requirements specification and testing: at every stage of development, experiments are based on *(a)* requirements identified earlier in the design process, *(b)* an assembly of components validated in a previous stage of testing, and *(c)* the appropriate type of test platform.

The conceptual difference between design and testing is easily obscured at early development stages. In (simulation-based) *design*, the focus is on structural and parametric changes to a (simulation) model, which lead to an incremental adaptation of a system design. In contrast, for *testing*, the system is fixed, and an experiment is set up to quantify a property or to validate a hypothesis (e.g., function, performance) about the present system design. As the system grows in scale and complexity, also the formulation of a test hypothesis becomes non-trivial; on one hand it is driven by the more complex system requirements, on the other hand also larger and more complex experimental setups are required. A holistic test description would support this re-framing from engineering design to test design, helping to narrow down the test purpose and test system requirements.

2.2 The Need for System Testing and Its Support

Section 2 outlined basic needs for system validation, and highlighted some of the existing approaches. The need for testing an integrated solution has been motivated in Sect. 2.2. In spite of different test realisations, there is common agreement that 'System testing' refers to testing at higher levels of system integration. With reference to Fig. 1, this notion of system testing thus refers mainly to the testing variants 'functional validation' and 'system validation'.

At the more basic levels, for components and sub-systems, requirements and test specifications are likely made by the developers themselves. For the higher levels, typically a long time passed between the initial formulation of functional and system requirements and the developed solution, increasing the gap between requirements and test execution.

Further, as outlined above, a wide variety of test platforms for multi-domain system testing are becoming available. Today, these test platforms have sufficient complexity of their own to concern the user with, rather than the object under investigation.

2.3 A Generic Procedure for System Validation

A procedural support can be useful when adopting a complex test platform attempting validation of a complex integrated control solution.

A holistic view on testing procedures is illustrated in Fig. 2. At the outset, this procedure template connects the system definition and use cases with a test objective in a test case. Once this link is fully established, the test specification captures fully the requirements for an experimental setup. The test platform can now be identified and suitably configured, even as a complex one that connects several research infrastructures (here: RI *a* and RI *b*). The experiment execution in the infrastructure and subsequent result evaluation may now lead to judging the test as successful, returning information with reference to the specifications and test case; or it may lead to a re-iteration of the specifications.

Depending on the kinds of test purposes, relevant test platforms, devices or systems under test, etc., different procedures and methodologies are applicable. Under the conceptual frame of this holistic test procedure, the ERIGrid project defined specific approaches within co-simulation, multi-RI experiments, and hardware-in-the-loop testing. For instance, a concrete test procedure, the "testing chain", as described in Sect. 2. To address the work with large-scale systems in a co-simulation context, an approach was formulated in [23], as reported in Sect. 4.

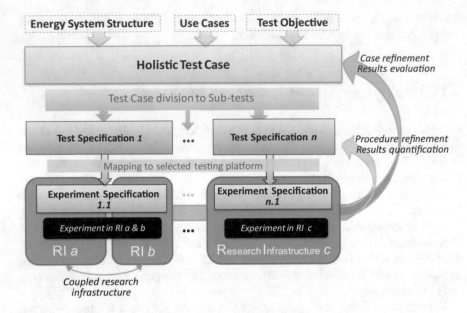

Fig. 2 Outline of the holistic test procedure with three research infrastructures, of which two are coupled

Finally, the holistic test description methodology, outlined in Chap. 3, offers systematic support for the formulation of concrete testing initiatives of any complexity and suits as semantic framework for further testing harmonization and test automation.

2.4 Testing Chain

The state-of-the-art in testing involves simulation, lab testing and field testing in that sequence. This testing approach lacks smooth transition and lacks coverage of smart grid functionalities. The "Testing chain" approach [4, 16], however, covers the whole range of testing possibilities including simulation, Software-In-the-Loop (SIL) [3], CHIL, Power Hardware-In-the-Loop (PHIL) and field testing sequentially. Such method can investigate the whole range of functions and hardware in the test system resulting in cost efficient validation. This kind of testing is composed of a series of tests with increasing complexity and realism. This is the general approach to follow for developing a new component or algorithm which affects system behaviour. The gradually increasing realism of the testing chain allows to develop a product in a relevant environment saving time and money. This could be suitable for the device manufacturers and software developers. Indeed, the first step of a developing phase is a pure simulation experiment; then, if the results are good, the object under test

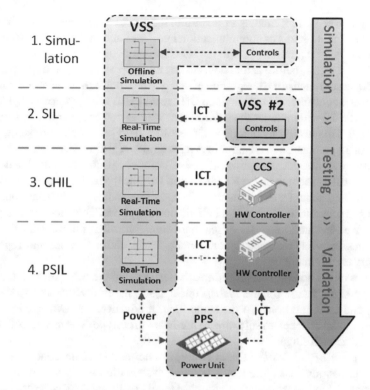

Fig. 3 Testing chain concept for CPES validation

is tested in a more relevant environment (from a CHIL experiment, where there is a real behaviour of the controller, to the PHIL, where the experiment takes into account the real behaviour of the whole OuI). All these steps are recommended before testing the OuI in the real environment. Testing products in environments with increasing complexity helps to identify and solve any critical aspect that could affect the performances.

Figure 3 provides an overview of the proposed testing chain. In Stage 1 investigations performed in a pure software simulated environment are usually carried out in steady state or transient conditions. This enables the functionality test of the control algorithm but does not represent adequately the interface between power and control systems.

Stage 2 of the testing chain proposes the use of two dedicated software tools for executing the power system model and controller separately. This SIL simulation or co-simulation technique allows the exchange of information in a closed loop configuration. After verifying the correct behaviour of the control algorithm in Stages 1 and 2, Stage 3 deals especially with the performance validation of the actual hardware controller by the use of a CHIL setup. CHIL testing provides significant benefits compared to simulation-only and SIL experiments. Using RT Simulator for executing power system models in real time, the actual hardware controller can be tested including all kinds of communication interfaces and potential analogue signal measurements by interfacing it with the RT Simulator.

The final Stage 4, before actual field-testing and implementation, of the proposed testing chain approach is the integration of real physical power hardware controlled by the hardware controller. This combined CHIL and PHIL is called Power System-in-the-Loop (PSIL) and includes the controller as well as power apparatus like inverter, motors, etc. This testing technique is closest to a field test of a component, which still can be implemented in a laboratory: it integrates real-time interactions between the hardware controller, the physical power component and the simulated power system test-case executed on the RT Simulator. Despite the high complexity to ensure stable, safe and accurate experiments, a PSIL setup enables an investigation, not as a single and separate entity, but as a holistic power system. This technique is proven to validate entire functionalities of real hardware controller, interdependencies and interactions between real power components in an entire flexible and repeatable laboratory environment.

In terms of the holistic test procedure, the testing chain realises several iterations, utilizing a static frame for the system under test, enabling efficient re-use of test systems and configurations. At the same time, it advances the OuI in each testing step from model concept, to software, to hardware prototype. The test objectives will be adapted at each stage.

Application of the Testing Chain concept: The testing chain concept has been adopted for a study case aiming at generating systematic improvements on the performance of a converter control function. Details of the test case descriptions are found in Sect. 4.1 and the full study case with results are reported in Sect. 4.

3 ERIGrid Holistic Test Description Methodology

This section introduces the developed formal concepts of test description, based on related work on testing and on requirements engineering in the energy systems domain. First relevant background is introduced, then the method is presented and finally exemplified based on application experiences.

3.1 The Requirements and Semantics of Test Description

Testing and experimentation with system solutions occurs in context of a design and engineering process, as outlined in Sect. 2.1. From this point of view, engineering requirements and their formalisation in shape of use cases and system configurations are one input to a holistic test description. The test execution itself requires a device or object(s) to be tested as well as the test platform.

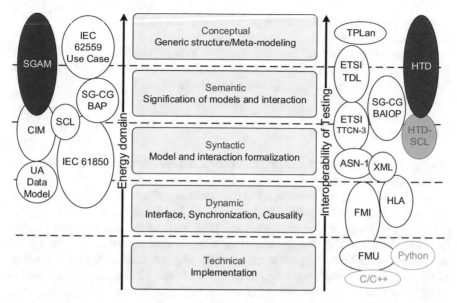

Fig. 4 Abstraction layers of a holistic test and the related standards

3.1.1 Energy System Semantics and Requirements

The existing energy system semantics (or information models) are presented on the left side of Fig. 4. The common information model (CIM/IEC61970-61968) [7, 8], OPC UA data model [21] and IEC 61850 data model [6] are widely utilized in the electrical domain. These standards address the functional, semantic, and syntactic configurations of a system. However, the technical and dynamic configurations are provided by the specific implementation technologies. Given these specific modelling standards for electric domain, describing and modelling the other domains such as ICT and thermodynamics in the system specifications requires further support. Nevertheless, the energy system semantics can be used as building blocks for the CPES design.

The Smart Grids Architecture Model (SGAM) proposes an interoperability architecture that covers mainly the conceptual and semantical interactions in a multi-domain smart grid. The link between SGAM and a validation is presented as a methodology based on use-case reference designation and specifications [1]. The SGAM methodology uses IEC 62559 for energy system design and provided the tailored use case template for this purpose. In this concept, a use case is considered as the basis for defining a system, its functionality and interaction necessary for the experiment design. It involves also the definition of Basic Application Profiles (BAP) and Basic Application Interoperability Profiles (BAIOP) as modular elements for the specification of system and subsystem. BAP and BAIOP represent the basic building blocks for the CPES and can provide possible skeletons for setting up interoperability validation experiment [19].

It is noteworthy that the use-case specifications provided in BAP and BAIOP involve specifically the system/sub-system architecture, but that they lack guidelines for the test specifications, implementation and technologies.

3.1.2 Testing Semantics

The link from the above information models and requirements to the validation setup is obscured, hence, the specification gap introduced in Sect. 2.1.

In the communication domain, ETSI provides a set of testing semantics including the ETSI test description suite, which consists of Test Description language (TDL) [14], the Test Purpose Language (TPLan) [13], and the Testing and Test Control Notation Version 3 (TTCN-3). While TPlan addresses the objective and scope of the test regardless to the testing environment, TDL fills the methodology gap between TPLan and the complex executable semantic. TDL and TPLan are mapped then to TTCN-3, where it specifies syntax, glossaries and templates to characterize a test configuration and procedure. However, still a corresponding test system is needed for the execution, i.e., the TTCN-3 semantic needs to be mapped down to an execution platform and can be integrated with system types of other languages (ASN.1, XML, C/C++). Besides, as a test specification semantic, TTCN-3 requires a domain specified syntax and vocabularies to enable comprehensive communication among its elements. The concept of abstract test suite in TTCN-3 standard [22] represents test descriptions in information technology. By defining formal (standardized) testing semantics and syntax, TTCN-3 enabled test automation [20], a software suit for conformance testing [24], and to promote reusability and possibility for further integration of new elements into the framework [5]. For instance, TPLan, TDL and TTCN-3 are utilized in information domain. However, in order to apply them to CPES assessment and validation, there is missing a means to establish a concrete link to energy system specifications, as the ETSI suite is not meant to interface physical structures and functions. This gap may be filled by integration of a complementing energy system semantic for testing.

The holistic test description addresses both energy system semantics and testing semantics, offering specification levels that relate to energy systems use cases and structural descriptions, while offering descriptions levels associated with a particular test platform. This multi-level specification is conceptually similar to those defined in the ETSI suite of TPLan, TDL, and TTCN-3.

3.2 The ERIGrid Test Description for System Validation

The holistic test description is a set of documents and graphical representations intended to support its users in the definition of complex tests. It follows the system

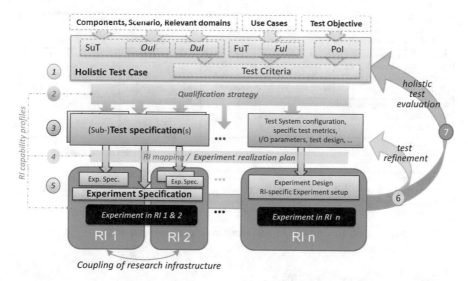

Fig. 5 The elements in context of the holistic test procedure

validation procedure outlined in Sect. 2.3. In that it can lead to a better planning of the experiments and help in the mapping of those experiments to different laboratories by contributing to clarify the test objectives, setups and parameters of interest. The whole process and some of its key concepts are illustrated in Fig. 5.

The three key levels of in test description are:

A **Test Case** provides a set of conditions under which a test can determine whether or how well a system, component or one of its aspects is working given its expected function.

A **Test Specification** defines the test system (i.e. how the object under investigation is to be embedded in a specific system under test), which parameters of the system will be varied and observed for the evaluation of the test objective, and in what manner the test is to be carried out (test design).

An **Experiment Specification** defines by what exact means a given test specification is to be realized in a given laboratory infrastructure.

From the practical perspective, the Holistic Test Description (HTD) is a set of templates for each level, and an associated graphical notation for system configurations. It constitutes a flexible framework that can be adapted according to the users' needs or the test cases to be applied. The steps of the holistic test procedure (Fig. 2) from the abstract conception of the experiment to the physical implementation in a laboratory are:

1. Test Case (TC)
2. Qualification Strategy (QS)
3. Test Specification (TS)
4. Experiment Realisation Plan

Test Objectives Why is the test needed? What do we expect to find out? A short narrative of context and goals of the test.		Purpose of Investigation *(PoI)* The test purposes classified in terms of *Characterization, Verification,* or *Validation.*	
Object under Investigation *(OuI)* "the component(s) (1..n) that are to be qualified by the test"	Function(s) under Investigation *(FuI)* "the referenced specification of a function realized (operationalized) by the object under investigation"	System under Test *(SuT)* Systems, subsystems, components included in the test case or test setup.	Functions under Test *(FuT)* Functions relevant to the operation of the system under test, including FuI and relevant interactions btw. OuI and SuT.
Domain under Investigation *(DuI)*: "the relevant domains or sub-domains of test parameters and connectivity."			
Test criteria: Formulation of criteria for each PoI based on properties of SuT; encompasses properties of test signals and output measures.			
target metrics Measures required to quantify each identified test criteria	variability attributes controllable or uncontrollable factors and the required variability; ref. to PoI.	quality attributes threshold levels for test result quality as well as pass/fail criteria.	

Fig. 6 Test case template as canvas [12] can be retrieved as download from [2]

5. Experiment Specification (ES)
6. Results Annotation
7. Experiment Evaluation

The TC template collects the motivation for the test. It frames the purpose of the test, the domains and sub-domains with their connectivity, the test setup, the relevant functions and the metrics to identify whether the test performed has been successful. The TC is an essential part of the testing effort as it represents the first clarification of the test objectives. For complex experiments, a single TC can have several linked TSs and ESs downstream. To support the early drafting process for test case development, the TC template is suggested to be filled in a Canvas format (see Fig. 6), which represents all components of the test case template in visual relation. The TC formulations typically go through several refinements between initial conception and final documentation of a testing campaign. Especially for complex test cases, it is common to break down a test objective into several PoIs and various Test Criteria for each PoI. It outlines how the OuI is going to be characterized or validated by means of a set of tests.

The next steps, the QS, TS, ERP all support the concretization and breaking down of a test case toward executable experiments. The QS is focused on describing this break-down, in a free-form textual description, but can also be represented as a graph expressing the hierarchical relation between TC and multiple different TSs and ESs. The TS addresses a specific PoI in detail and defines a concrete test system configuration, the test design, the parameterization, metrics and test sequences. The Experiment Realization Plan aims to identify at which particular RIs the respective TSs could be implemented in terms of hardware, software, models. Up to this moment, the methodology assumes that the description of the TC is independent of the RI. In practice, this assumption is not always valid, so that information from the laboratory can influence e.g. the acceptable complexity of a TS. The ES defines the mapping of the TS to the components, structure and procedures of a laboratory. As it is required to know many details about the components, measuring devices, expected uncertainties, etc. it should be prepared in collaboration between a technical manager of the RI and the user.

From experience, information the laboratories available to external users is typically insufficient to plan an experiment without the involvement of local experts. Here the HTD approach can be particularly helpful in facilitating the communication between external users and laboratory staff. As a guideline, the external user should be 'owner' of the steps from TC to TS. The local staff however, should 'own' the ES, to ensure a feasibility and integrity of the experiment design with laboratory capabilities.

The last two steps of the procedure, Results Annotation and Experiment Evaluation are not subject to the HTD framework. The process of registering the results of the tests depends on the test itself and the only advice given to the users is to keep them traceable among the different test platforms, time resolutions and data formats. A method for exchangeable file formats and annotation of experiment result data is found in [10]. The results obtained in the testing process provide feedback for the clarification of the TS. The final evaluation of the conducted experiments serves as input for the refinement of the holistic TC.

3.3 Holistic Test Description: Key Concepts

A comprehensive framework for test description requires the introduction of a few concepts and their definition to contrast with the blurry lines of their everyday use.

Test objective is the purpose for carrying out the test. These can be divided into three categories:

- Characterization test: a measure is given without specific requirements for passing the test. Examples: characterizing performance of a system; developing a simulation model.
- Validation test: functional requirements and abstract measures are provided but are subject to interpretation; qualitative test criteria. Example: is a controller ready for deployment?
- Verification test: Tests where requirements are formulated as quantitative measures and thresh-olds of acceptable values are quantified. Example: Testing if a component conforms to a standard.

3.3.1 System Configurations in Test Descriptions

System configurations, use cases and test cases form a logical chain that can be applied throughout a development project. The main concepts are as follows:

System(s) Configuration is an assembly of (sub-)systems, components, connections, domains, and attributes relevant to a particular test case.

Table 1 Overview of system configuration levels [12]

SC type	Generic SC	Specific SC	Experiment SC
Described in	Test case	Test specification	Experiment specification
Topology	Domain-coupling	SuT components	Testbed and OuI
Parameters	NO	Partial, preferred values	YES
OuI concrete	NO	YES	YES
Non-OuI concrete	NO	NO	YES

A **Component** is constituent part of a system which cannot be divided into smaller parts without losing its particular function for the purpose of investigation Remark: In a system configuration, components cannot further be divided; connections are established between components.

A **System** is defined by a system-boundary, and can be composed of sub-systems, or components that cannot be further decomposed in the relevant context. It is described as a set of interrelated elements considered in a defined context as a whole and separated from their environment. Remark: In a system configuration, a system, which may be divided into sub-systems, represents a grouping of components; functionality can be attributed to systems and components and vice versa.

Connections defines how and where components are connected, and connections are associated with a domain.

Domain is an area of knowledge or activity in the context of smart grids characterized by a set of concepts and terminology understood by practitioners in that area, typically infrastructure-specific operation areas such as electricity, heat, primary energy resources or ICT. Multi-domain components thus act as interface (conversion) between domains. Finally, a domain can be divided into sub-domains.

Attributes define the characteristics of components and systems, such as parameters and states.

Constraints describe limitations of component attributes, systems, domains or functionality.

For each layer of test description a different interpretation of the system configuration is relevant, as illustrated in Table 1 and Fig. 7. Table 1 provides an overview of the differences between the different SCs.

As an example of the three levels, Fig. 7 shows system configurations from a test involving coordinated voltage control of remotely controllable Photovoltaic (PV) inverters.

In the GSC, only coupling domains and high-level subsystems are specified, while the number of units involved is not specified. The Test System (SSC) identifies the OuI as a single inverter but requires both the coordinated voltage controller and several other inverters to be connected to a distribution system. Finally, in the experiment setup (ESC), elements required to emulate signals for the OuI are specified. These, together with a specification sheet (not shown), serve as a complete documentation

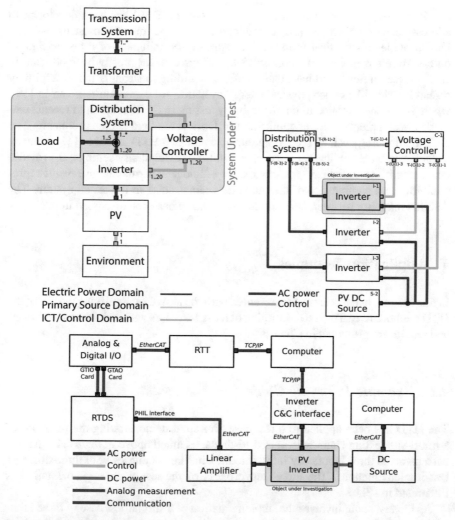

Fig. 7 System configurations for a coordinated voltage control test case [11, 12]: generic system configuration, specific system configuration, experiment system configuration

of the experimental setup. Only one PV inverter is seen in a PHIL setup, while the voltage controller is implemented in software, while the other inverters as well as the distribution grid are simulated on a digital real-time simulator.

3.4 Remarks on Quantitative Assessment

As most testing is quantitative, also a framework for quantitative selection of test parameters and result evaluation is needed.

The statistical concept of *Design of Experiments* (DoE) has been developed to address result significance and reproducibility in experimentation. In its essence, DoE provides a statistical framework to explore the influence of different *factors* on a system's *response*. The concepts of DoE have so far mostly been adopted in single research fields and have had a difficult standing in strongly interdisciplinary research fields like cyber-physical energy systems. An exception is given in [19], where it has been applied to interoperability testing in CPES relation to recent standards developments. Further application of DoE in the field is thus promising. The DoE methodology can be seen as an intrinsic part of a HTD, and the HTD is meant to facilitate DoE application in complex and interdisciplinary settings. It provides testing with the statistical groundwork for efficient experimentation, result reproducibility and significance of the outcome against noise in the tested system. The detailed mapping between DoE and holistic testing has been discussed in [18].

4 Application Examples

In this section, two test cases are presented with the aim of exemplifying the use of the HTD methodology. The full description of these test cases with their implementations and results are presented in Chap. 6.

4.1 Example 1: Testing Chain

The HTD has been applied to a test case aiming at demonstrating the potential of a multi-site testing chain with varied testbeds, as noted above in Sect. 2.4. The test case involves three laboratory infrastructures in three countries with three different test implementations. The three-step process of the test-chain implementation is illustrated in (Fig. 8).

As the test chain involves the implementation of a similar test case in three laboratories and also due to the need for model and results exchange among the involved laboratories, harmonized specification of the test case with unified template was crucial. The utilization of the HTD methodology in this test case involved three stages: jointly specifying of the test case and the qualification strategies, specification of the test by partners with common purposes of investigation and finally the specification of the experiments by the individual laboratory infrastructures. Short version of the test case description of this test is presented below:

- Name of the Test Case: Testing of converter controller through multi-site testing chain with varied testbeds
- Narrative: This test case aims at demonstrating the potential of a multi-site testing chain with varied testbeds for generating systematic improvements on the performance of a converter control function.

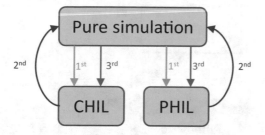

Step 1: Run simulation and prepare load profiles for consequent tests

Step 2: CHIL and PHIL results with recommendation for improvement of converter controller

Step 3: Re-run Simulation, CHIL and PHIL tests to validate controller improvements

Fig. 8 Test-chain implementation process

- Function Under Investigation: Converter RMS controller (receiving P/Q setpoints and setting d/q axis current setpoints)
- Object Under Investigation: Converter RMS controller subsystem
- Domain Under Investigation: Electric power domain, Control domain Purpose of Investigation:

  - PoI 1: Characterization of converter controller influence of the system performance.
  - PoI 2: Validation of model exchange among RIs.
  - PoI 3: Validate improved control system performance.

- System Under Test—illustrated in Fig. 9:
- Function Under Test: Converter Q/V and P/f controller algorithm, inner current controller, a low voltage distribution grid connecting five loads, four PV and a battery.
- Test Criteria: Settling time, overshoot, damping factor and peak time for a step response after step changes of PV output and the load connected with the PV.

  The testing campaign was carried out as well and is reported in Sect. 6.

4.2 Example 2: Coordinated Voltage Control

Another example case can be the Coordinated Voltage Control (CVC) case involving flexibility from DER, communication infrastructure and centralized optimized control. The related use case was introduced in Sect. 2.3. To specify this test case, the three level specification templates of the HTD are applied detailing the test from

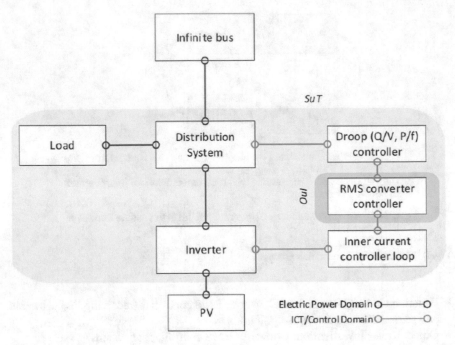

Fig. 9 Test-chain system under test

generic to specific to laboratory level plans. The three level specifications are associated with test case, test specification and experiment specification respectively. The
main questions addressed in the test case template are why the test is needed and what
the objectives are. Some of the main specifications of the CVC test are presented
below.

- The Narrative: For a Distribution Management System (DMS) Voltage controller
  in development stage (simple implementation) the performance of the DMS algorithm and controller should be evaluated under realistic conditions. This test could
  be seen as the last step before installing the DMS in the field.
- The system under test: includes DMS, DER, OLTC, transformer, distribution lines,
  telecom network as shown in (Fig. 1).

  – Object under investigation: DMS controller.
  – Domain under investigation: Electric power, ICT.

- Function under test: includes DER P,Q control, measurements, OLTC tap control,
  communication via ICT.

  – Function under investigation: optimization in the controller, state estimation.

- Test objectives/PoI: Characterization and validation of the DMS controller.

  – Convergence of the optimization (validation).

- Performance of the optimization under realistic. conditions (characterization)
- Accuracy of the state estimation (characterization).

- Target measures:

  - 1. convergence (when/how often?), 2. How fast?, 3. solution quality.
  - Voltage deviation of all the nodes from 1 pu, number of tap changes, network losses.
  - Voltage, P, Q estimation errors.

- Variability attributes: Load patterns (realistic, annual variation); Communication attributes (packet loss, delays).
- Quality attributes (thresholds): 1.2: convergence within 2 sec (validation), 3.* estimation quality characterized with confidence 95%.

After the CVC test case is described with the details of purposes of investigation, logical break down of the test case into sub-tests follow. In this process, a strategy will be developed identifying testbeds targeting to meet the requirements of the test case. Finally, in the experiment specification details of the components to be used in the test, such as type and ratings of OLTC and DER, are specified. Furthermore, the connectivity of the components and also the actual values of the variability attributes, such as load patterns, are specified. The full specification of the CVC test can be found in [15]. A detailed similar test case is provided in Chap. 6.

5 Conclusion

As advanced testing platforms are becoming part of a multi-disciplinary development process, also testing campaigns need to integrate information of multiple viewpoints. To support the planning and documentation, this chapter presented a model and method for detailed test planning that is suitable for even complex test campaigns. This method, called 'holistic test description', relates system requirements to test design and testing platforms; it complements the analytical design of experiments with a test engineering process. For further details and instructions on the described method, please refer to [12]. Templates, guidelines and further examples are also found on the corresponding GitHub site.[1]

With the testing chain, a prototypical process for integrated multi-stage system development validation was introduced. An abstract testing process was outlined, so that the presented tools for handling the information between system requirements and test platform configuration. This chapter has illustrated how the management of testing campaigns can be supported by a structured approach on information management and the systematic use of advanced test platforms.

---

[1] https://github.com/ERIGrid/Holistic-Test-Description.

Going forward, further advanced testing will be introduced. Then, in Chap. 6 two example testing campaigns are reported, and an overall evaluation of the here introduced test description method is summarised in Chap. 7.

References

1. Methodologies to facilitate Smart Grid system interoperability through standardization, system design and testing. Smart Grid Mandate CEN-CENELEC-ETSI Smart Grid Coordination Group, Tech. Rep., (2014)
2. Holistic test description templates, ERIGrid (2019). https://github.com/ERIGrid/Holistic-Test-Description
3. Babazadeh, D., Chenine, M., Zhu, K., Nordström, L., Al-Hammouri, A.: A platform for wide area monitoring and control system ICT analysis and development. In: 2013 IEEE Grenoble Conference, pp. 1–7 (2013)
4. Brandl, R., Kotsampopoulos, P., Lauss, G., Maniatopoulos, M., et al.: Advanced testing chain supporting the validation of smart grid systems and technologies. In: 2018 IEEE Workshop on Complexity in Engineering (COMPENG), pp. 1–6. IEEE (2018)
5. Broy, M., Jonsson, B., Katoen, J.P., Leucker, M., Pretschner, A.: Model-Based Testing of Reactive Systems: Advanced Lectures (Lecture Notes in Computer Science). Springer, Berlin (2005)
6. International Electrotechnical Commission: IEC61850 - Power Utility Automation. Technical report TC 57 - power system management and associated information exchange (2003)
7. International Electrotechnical Commission: Application integration at electric utilities - System interfaces for distribution management - Part 11: Common information model (CIM) extensions for distribution. Technical report, TC 57 - Power system management and associated information exchange (2013)
8. International Electrotechnical Commission: Energy management system application program interface (EMS-API) - Part 301: Common information model (CIM) base. Technical report, TC 57 - Power system management and associated information exchange (2014)
9. Forsberg, K., Mooz, H.: System engineering for faster, cheaper, better. In: INCOSE International Symposium, vol. 9, pp. 924–932. Wiley Online Library (1999)
10. Gehrke, O., Jensen, T.: Definition of a common data format. Deliverable, SmILES Consortium (2018)
11. Heussen, K., Morales Bondy, D.E., Nguyen, V.H., Blank, M., et al.: D-NA5.1 Smart Grid configuration validation scenario description method. Deliverable D5.1, ERIGrid Consortium (2017)
12. Heussen, K., Steinbrink, C., Abdulhadi, I.F., Nguyen, V.H. et al.: ERIGrid holistic test description for validating cyber-physical energy systems. Energies **12**(14) (2019)
13. European Telecommunications Standards Institute: Methods for testing and specification (mts); tplan: A notation for expressing test purposes. Technical report, ETSI ES 202 553 V1.2.1 (2009)
14. European Telecommunications Standards Institute: ETSI test description language. Technical report (2018). https://tdl.etsi.org
15. Kotsampopoulos, P., Maniatopoulos, M., Tekelis, G., Kouveliotis-Lysikatos, I., et al.: D-NA4.2a Training/education material and organization of webinars. Deliverable D4.3, ERIGrid Consortium (2018)
16. Maniatopoulos, M., Lagos, D., Kotsampopoulos, P., Hatziargyriou, N.: Combined control and power hardware in-the-loop simulation for testing smart grid control algorithms. IET Gener. Trans. Distrib. **11**(12), 3009–3018 (2017)
17. van der Meer, A.A., Palensky, P., Heussen, K., Bondy, D.E.M., et al.: Cyber-physical energy systems modeling, test specification, and co-simulation based testing. In: 2017 Workshop on Modeling and Simulation of Cyber-Physical Energy Systems (MSCPES), pp. 1–9 (2017)

18. van der Meer, A.A., Steinbrink, C., Heussen, K., Morales Bondy, D.E., et al.: Design of experiments aided holistic testing of cyber-physical energy systems. In: 2018 Workshop on Modeling and Simulation of Cyber-Physical Energy Systems (MSCPES), pp. 1–7. IEEE (2018)
19. Papaioannou, I., Kotsakis, E., Masera, M.: Smart grid interoperability testing methodology: a unified approach towards a European framework for developing interoperability testing specifications. In: EAI International Conference on Smart Cities Interoperability and Standardization (2017)
20. Schieferdecker, I.: Test automation with TTCN-3-state of the art and a future perspective. In: IFIP International Conference on Testing Software and Systems, pp. 1–14. Springer, Berlin (2010)
21. TC 65/SC, E.: IEC 62541-1: OPC unified architecture - part 1: Overview and concepts. Technical report, International Electrotechnical Commission (IEC) (2010)
22. ETSI Centre for Testing and Interoperability: TTCN-3 tutorial. Technical report, (2013), available at: http://www.ttcn-3.org/files/ETSI_TTCN3_Tutorial.pdf, accessed 17.04.2020
23. Widl, E., Spiegel, M., Findrik, M., Bajraktari, A., et al.: D-JRA2.2 Smart Grid ICT assessment method. Deliverable D8.2, ERIGrid Consortium (2018)
24. Zeiss, B., Kovacs, A., Pakulin, N., Stanca-Kaposta, B.: A conformance test suite for ttcn-3 tools. Int. J. Softw. Tools Technol. Trans. **16**(3), 285–294 (2014)

Simulation-Based Assessment Methods

A. A. van der Meer, R. Bhandia, P. Palensky, M. Cvetković, E. Widl, V. H. Nguyen, Q. T. Tran, and K. Heussen

1 Introduction to Smart Grid Modelling and Simulation

In general, smart grids can be considered as the application of various types of automation and control for the operation of energy technology, with a focus on electrical power engineering. Examples include the application of distributed automation in substations, smart metering of domestic consumers, and wide-area protection mechanisms. Such technology allows the energy systems to be operated and controlled more optimally and to be pushed to their design boundaries. Notwithstanding these advantages, these concepts heavily rely on ICT structures, which form the glue between the physical domain (e.g., energy systems and their components) and the control and automation domain (e.g., decision-making devices, overarching logic and algorithms). Together, these domains constitute the concept of cyber-physical energy systems (CPES).

The domain coupling challenges of CPES are evident. The overall system exhibits multi-time scale (transients versus market decisions) interactions, multi-size (decentralised measurements, wide-area protection) properties, and heterogeneous (physical versus discrete events) behaviour. In order to assess the operation, security, and reliability of CPES, the common way of testing and validation for smart energy components shall be reconsidered. Eventually, lab-based approaches to test, validate

A. A. van der Meer (✉) · R. Bhandia · P. Palensky · M. Cvetković
Delft University of Technology, Delft, The Netherlands
e-mail: a.a.vandermeer@tudelft.nl

E. Widl
AIT Austrian Institute of Technology, Vienna, Austria

V. H. Nguyen · Q. T. Tran
Université Grenoble Alpes, INES, Le Bourget du Lac, France

CEA, LITEN, Le Bourget du Lac, France

K. Heussen
Technical University of Denmark, Roskilde, Denmark

© The Author(s) 2020
T. I. Strasser et al. (eds.), *European Guide to Power System Testing*,
https://doi.org/10.1007/978-3-030-42274-5_3

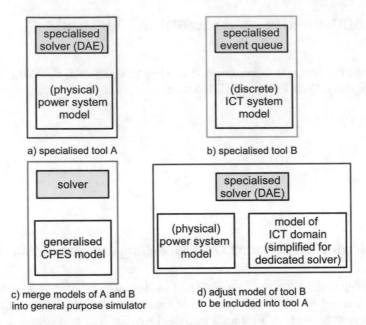

Fig. 1 Overview of simulation based assessment of CPES. Domain-specific approaches (i.e., **a** and **b**), and multi-domain simulations (i.e., **c** and **d**)

and roll-out new concepts must be able to capture the cross-domain interactions the system or component under test will be subject to.

One of the steps that need to be taken to achieve this is analysis, modelling, and simulation of cross-domain interactions. Figure 1a and b show for example of how domain-specific models (e.g., physical and ICT) are usually simulated by dedicated tools with specialised solvers. Cross domain interactions can be included by attempting to model the entire system under test in a general-purpose simulation tool like Simulink or OpenModelica (i.e., Fig. 1). This approach has the advantage of maintaining the entire model into one simulation tool. A common downside it that such models commonly scale badly in size and phenomena addressed. Another method is to include the model of the 'alien' domain (say model B of Fig. 1 d)) into a specialised tool and subsequently make this model compatible with the applied solver. This is usually done when it is assumed justifiable to simplify parts of the overall model to make it suitable for a single-domain tool.

Figure 1 often lead to a suboptimal trade-off between simulation efficiency (speed) and accuracy of the phenomena of interest. This chapter will focus on the simulation aspects of this challenge and more specifically on one particular method to deal with this: coupled simulations, also referred to as co-simulations [7]. As an assessment approach, co-simulation offers key advantages for the simulation of cyber-physical systems-of-systems:

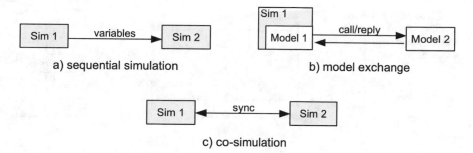

Fig. 2 Variants of coupled simulations

1. *Modularity*—Co-simulation allows to represent (parts of) sub-systems with the most appropriate tool available. This encourages a modular representation of the system under test, with clean semantic and functional model boundaries along the real-world domain borders.
2. *Hierarchical composition* as a feature of system-of-systems architectures, is supported in co-simulation by the modular approach, where a hierarchical modelling strategy allows to switch out abstracted functional representations with explicit models of system layers (e.g., abstracted ICT layer: point-to-point information exchange, detailed: explicit transport layer model).

In the following, the basic concepts of co-simulations will be explained. Then the available standardized approaches to set up a co-simulation, such as the high-level architecture and the functional mock-up interface will be introduced. Finally, a survey of scaling aspects in terms of co-simulation of CPES follows and an example implementation is discussed, which concerns coupling a power system simulation to a general-purpose simulation.

2 Co-simulation Based Assessment

2.1 Introduction to Co-simulation, Goals, and Challenges

The term co-simulation is typically used, when two or more models are used in one simulation. (Real) Co-simulation happens, when these two or more separate models are executed concurrently and if their variables or states depend on each other (i.e., option c in Fig. 2). These simulators have to synchronize with each other periodically. Sometimes embedding a model into another one and executing them with just one simulator (i.e. numerical solver) is called co-simulation with model exchange. If one simulation component is just uni-directionally using data (e.g., time series) from another simulation we speak of sequential simulation.

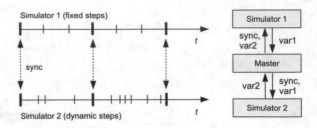

Fig. 3 A master algorithm synchronizes the simulators and passes on shared variables

Figure 3 shows the time sequence and data flow of two simulators that are coupled via a master algorithm. The master must have some possibility to start and stop the simulators, ideally in an on-the-fly fashion that does not require re-initialization of the states. The choice of synchronization steps is usually up to the co-simulation engineer. If both simulators have fixed time steps, and if these time steps are multiples of each other, the synchronization becomes easy. If, however, the time steps are totally independent of each other the master might have to interpolate variables that are exchanged between two-time steps. This situation in shown in Fig. 3 and might require the master to roll-back simulators from time to time if possible and required.

The reasons why co-simulation is often used are usually pragmatic and solution oriented:

- Existing legacy models can be used with new models. Often it is not feasible to re-implement existing models in the simulator of choice with the given resources (such as shown in Fig. 2b).
- Specialized simulators can be used for parts of a multi-disciplinary problem. By that, models of one sub-model (e.g., discrete events) do not have to be badly "imitated" in the simulator of another sub-problem (e.g., continuous dynamics). The specialized simulators usually have a better (and maybe even validated) model library and a tailored work-flow and user-interface for their particular domain.
- The simulation study can have multiple foci. Unlike in the case of a standard (monolithic) simulation, there is no need to simplify sub-problems. Each sub-problem (may it be mechanical, thermal, electric, economic, etc.) can be modelled in all detail since it runs (and can even be tested) in its own specialized environment.

These advantages have to be contrasted with a number of disadvantages, too:

- Models in different modelling environments need to be maintained. This involves multiple modelling languages, simulation project files, and software licenses. This also means that the staff, doing the simulation, needs to be educated in all these tools—plus the co-simulation environment! A high level of versioning and documentation discipline is required to achieve a sustainable way of working with that.
- Experience shows, that co-simulation is slow. Although its perfect suitability for parallel computation would suggest speed gains, it is the synchronization of simulations (that were often not designed to be synchronized) that slows things down. Some legacy simulators process licensing information when they are stopped and

restarted or require a fresh initialization, which of course grinds down performance if frequent synchronization is needed.

- Error propagation and estimation of co-simulation is poorly understood. The choice of simulation step sizes or synchronization points is therefore not trivial and is still subject of technical developments.

Interfacing with the master can be done via various APIs (application program interfaces), one that received broad industry support is called FMI: the Functional Mockup Interface. It is an open standard, based on a C-interface that offers the specification of required functions such as start, stop, step, synchronization, variable exchange, etc.

The master algorithm itself is in its core often very simple but associated functionality (such as scenario handling, data logging, distributed computing, etc.) can be quite complex. Again, there are a few popular master platforms, two of them being HLA (the high-level architecture) and mosaik.

Once the master and the simulators are set up, the work flow is very much as a standard simulation-based analysis: Scenarios are generated (e.g., parameter sweeps, etc.), and an optimizer or engineer runs these scenarios in a number of simulations until the expected result is found.

2.2 Current Co-simulation Standards and Their Functionality

The High Level Architecture (HLA) was originally developed under the umbrella of the Department of Defense of the USA in order to serve its high demands for a versatile simulation environment. The development was initiated in the early nineties, and the current HLA version is standardized under HLA 1516–2010 (known also as HLA Evolved) [1]. This standard does not focus on the implementation of the co-simulation master, but instead, establishes the list of services that must be provided by the master (in HLA terms called Run-Time Infrastructure (RTI)). Some of the greatest advantages of HLA are its versatility and configurability, while at the same time, these features amount to a steep learning curve for the co-simulation engineer.

The Functional Mockup Interface (FMI) was created to ease model exchange between vendors of various components assembling larger physical systems (one such example is automotive industry). Therefore, its primary focus is on model encapsulation (within so-called Functional Mockup Unit (FMU)) and its current standard, FMI 2.0, provides a comprehensive interface for model engagement (such as model evaluation, Jacobian retrieval, etc.) [4]. As the second step in its evolution, FMI was enhanced with a co-simulation interface (such as starting, stopping, stepping of the models). The current standard anticipates packaging of FMUs with and without internal solvers. If packaged with internal solvers, the FMUs can be directly included in co-simulation. Otherwise an external solver must be engaged to step the model within FMU.

FMI and HLA are complementary in nature, since FMI focuses on the engagement of models, while HLA focuses on the master services [5]. Today, many engineering simulation tools allow to export the models as FMUs, which improves the breadth of co-simulation scope.

Finally, mosaik was created with a particular intention to serve as a smart grid co-simulation framework, and as such, it is largely in tune with energy system applications [8]. In contrast to the previously mentioned standards, mosaik is a direct implementation of a master algorithm, and not a standard per se. It is written in Python, based on a discrete event scheduler and is capable of FMU integration. Besides FMU integration, it also provides interfaces for several common tools in the energy system realm (such as DigSilent PowerFactory, PandaPower, etc.). Since its primary user group are energy engineers, the elaborate co-simulation settings are greatly simplified, which represents mosaik's greatest advantage. A comparison of mosaik and HLA for a co-simulation of a power system control action is performed in [9].

3 Co-simulation Framework for Smart-Grid Assessment

3.1 Co-simulation Interfaces Based on FMI

In order to accurately simulate Smart Grids, the interaction between the domains of *electrical power systems*, *communication* and *automation and control* is of crucial importance. As a proof-of-concept, co-simulation interfaces based on the FMI standard have been developed for selected state-of-the-art tools, examples of which are described in the following.

3.1.1 Power System Simulation with PowerFactory

DIgSILENT PowerFactory[1] is a commercial tool for power system design and analyses. PowerFactory does not officially provide an FMI-compliant co-simulation interface. However, it provides an API that enables basic interactions with simulation models at run-time like setting/retrieving variables and calculating power flows.

Furthermore, PowerFactory provides the possibility to issue so-called *events* during time-domain simulations (more specifically, *RMS simulations* in PowerFactory) that can change the system state at a specified point in simulation time. This mechanism has been utilized to enable a dynamic interaction with simulation models at run-time. It is suited for co-simulation and has been integrated into a stand-alone FMU exporter tool.[2]

---

[1] DIgSILENT PowerFactory, http://www.digsilent.com, accessed April 17, 2020.
[2] The FMI++ PowerFactory FMU Export Utility, http://powerfactory-fmu.sourceforge.net, accessed April 17, 2020.

3.1.2 Communication Network Simulation with Ns-3

In recent years, *ns-3* has become very popular in the network simulation community. ns-3 is a highly flexible simulation package, which allows programmers to add new attributes without modifying the core of the source code, or having to deal with a specific, restricted and complex API. The default version of ns-3 comes with an extensive library of models, which can be used to describe the components and other aspects of communication networks (e.g., devices, channels, interfaces, protocols).

A dedicated ns-3 package called *fmi-export* has been developed, which provides all functionality needed for creating an FMU for C—Simulation from a user-defined ns-3 application (typically referred to as *script*). An FMU created with the help of this package implements a tool coupling mechanism that allows to control the execution of the ns-3 simulator and to establish a connection for data exchange during run-time.

3.1.3 Control Simulation with MATLAB

Despite the popularity and widespread use of the numerical computing environment *MATLAB*, there is so far only comparably little support within the context of FMI. For instance, the *Modelon FMI Toolbox*[3] and the *FMI Kit for Simulink*[4] offer the export of Simulink models as FMUs for Model Exchange, but so far there is no tool available that allows to provide MATLAB's full functionality via an FMI-compliant co-simulation interface.

Therefore, the *FMI++ MATLAB Toolbox*[5] has been implemented that provides two components: a *front-end component* to be used by the co-simulation master and a *back-end component* to be used by MATLAB. The corresponding interfaces are tailored to suit the requirements of the FMI specification and they implement the necessary functionality required for a master-slave concept, i.e., synchronization mechanisms and exchange of data.

3.2 Mosaik for Scenario Development and Simulation Orchestration

The *mosaik*[6] framework is an easy-to-deploy software package that facilitates the integration of new simulators as well as the creation of co-simulation experiments. This is achieved via a lightweight software core based purely on Python, a special

---

[3]FMI Toolbox for MATLAB/Simulink, https://www.modelon.com/products-services/modelon-deployment-suite/fmi-toolbox, accessed April 17, 2020.

[4]FMI Kit, https://www.3ds.com/products-services/catia/products/dymola/fmi/, accessed April 17, 2020.

[5]The FMI++ MATLAB Toolbox, http://matlab-fmu.sourceforge.net, accessed April 17, 2020.

[6]The mosaik Smart Grid co-simulation framework, http://mosaik.offis.de/, accessed April 17, 2020.

Component-API for simulator integration, and a Scenario-API for flexible simulator coupling. The mosaik framework is still under active development and new features are being introduced based on activities within the smart grid testing and validation community.

3.2.1 FMI Support

As an example, the *FMI++ Python Interface*[7] and the mosaik framework have been successfully combined for the co-simulation of FMUs. Several examples of importing FMUs have been implemented using the FMI++ Python Interface to interact with the FMU and mosaik's high-level component API to integrate it into the co-simulation. For this, especially the functionality for conveniently handling FMUs was extensively used, such as extracting the FMU, parsing its model description or the ability to refer to input/output variables by name (rather than the numerical value reference associated to each variable).

3.2.2 Handling of Cyclic Dependencies

The term *cyclic dependencies* refers to a co-simulation setup in which two (or more) simulators require data from each other to advance their state in time (i.e., Fig. 2c). These data dependencies may lead to deadlocks with all simulators waiting for data from each other, halting the whole simulation process. Therefore, proper handling of these cyclic dependencies is one of the most crucial tasks in co-simulation. This is especially true in the case of Smart Grid applications, which typically involve feedback loops and a strong physical coupling between the individual components and subsystems.

The co-simulation framework mosaik has been developed with a strong focus on flexibility in terms of configuring the connected simulators. Accordingly, the scheduling algorithm of mosaik is designed in a way to allow integration of any number of simulators. Furthermore, all integrated simulators may display different step sizes and even vary their step size over time. In order to guarantee the absence of deadlocks for any given setup, the handling of cyclic dependencies in mosaik has so far had some limiting characteristics. In particular, using mosaik's intuitive connection capabilities to establish cyclic data exchange between two or more simulators has been prohibited. Instead, users had to extend the simulator interfaces to realize cyclic data exchange, which obviously decreases the usability of mosaik for researchers with limited programming experience. Furthermore, the described solution in mosaik only supports serial data exchange schemes.

Recently, the capabilities of mosaik have been extended to allow for higher usability in the handling of cyclic dependencies. The basic idea of this extension is the separation of data exchange into two stages: Simulators may receive data either

---

[7]The FMI++ Python Interface, https://pypi.org/project/fmipp/, accessed April 17, 2020.

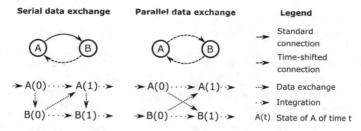

Fig. 4 Possible data exchange schemes in mosaik

before they are called to calculate a time step, or after they have calculated so that they store the data for the next time they are called. With this separation, priorities between simulators can be established so that deadlocks are avoided. Figure 4 illustrates different data exchange options between two simulators A and B. Connections for data exchange before calculations are called standard connections since they are part of the typical functionality of mosaik. The newly added connection type is called time-shifted connection since they provide data to simulators that already have been called for calculation.

Figure 4 shows that standard connections in mosaik provide data to a simulator for its calculation of the current time step while time-shifted connections provide data for the next time step to be calculated. Furthermore, mosaik provides the option to set default input data for the first calculation of a simulator that is addressed by time-shifted connections. In this way, parallel data exchange schemes may also be realized if initial input data can be assigned to each simulator. Overall, the extension of mosaik improves its usability and provides it with the most common options for handling cyclic dependencies in black box co-simulation for smart grid applications.

4 Scaling Considerations

The purpose of simulation-based smart grid assessment is often the ability to evaluate the (often non-linear) effects of changes in system parameters on large systems that cannot be established through abstracted analytical models or limited physical experiments with few hardware components.

Scaling-up of established simulation components to a large-scale scenario is conceptually simple in co-simulation: due to the modularity and hierarchical build-up of models, system components can be re-used with alternative parameters and scenario APIs allow scripted scenario configuration and handling.

However, there is a number of non-trivial issues that needs to be considered when planning and developing scale-up simulations, arising from either a) the complexity of system interactions represented, or b) the increasing simulation program scale and complexity [2]. Table 1 offers a view on several types of *large-scale phenomena* in energy systems, distinguishing whether these emerge from the physical domain

Table 1 Large scale phenomena considered in the context of smart grids

Real world (Investigated phenomena)	Physical (Laboratory)	Virtual (Simulation)
Scale in number of nodes and components	Number of nodes/buses/components	Number of equations
Complexity through inter-dependencies across multiple domains	Number of domains (power, heat, ICT etc.)	Number of simulation tools and instances
Complexity through stake-holder interpretations	Number of relevant layers (business, information, communication, components etc.)	Variety of models of computation (time-continuous, event-driven, stochastic, etc.)
Socio-geographical size	Geographical size	

Table 2 LSS phenomena characterization chart

Dependency on control parameters	Scale with the system size (linear, logarithmic, exponential, and polynomial)
	Appear at certain critical system sizes (i.e. phenomenon appears and remains beyond a certain control parameter value)
	Appear and disappear at certain operational zone of control parameters or parameter combinations
Variation of observation parameters	Extreme values (e.g. performance increase or decay; system failure)
	Inadvertent oscillations
	Intermittent performance degradation

(real-world application) or inaccuracies of the research infrastructure (laboratory or simulation environment).

To distinguish Large Scale System (LSS) phenomena, we characterize them by their effects on system parameters, as presented in Table 2. The phenomena are characterized by the observable relation between system input and control parameters—factors in design of experiment (DoE) terms—and the resulting variation of observation parameters (observables, performance metrics, DoE: response variables).

In order to consider appropriate assessment methods for the aforementioned categories, two principal scaling approaches can be adopted:

- *Upscaling in terms of system properties (i.e., scale out)*: this method targets phenomena directly related to physically large scale systems. E.g., how does the co-simulation scale with physical system size?
- *Upscaling in terms of simulation and modelling (i.e., scale up):* this method targets large scale implementations in models and simulation for the validation of smart

grids. E.g., how does the co-simulation scale with the number of models and simulations involved?

The co-simulation example demonstrated in the next section introduces a co-simulation that has been subject to upscaling principles: in terms of properties—scale out (rate power, number of wind turbines) and in terms of modelling—scale up (number of FMUs) [3].

5 Fault Ride-Through of a Wind Park Example

This section comprises a typical test case in which domains and their co-simulation challenges come together, with a focus is on the evaluation of cyclic dependencies between different models in the context of co-simulation. In the implementation example a standard IEEE 9-bus dynamic test system is modified to contain a Wind Power Plant (WPP), which replaces one of the 3 main generators. A wind park is typically subject to grid connection requirements by the network owner, formulated in grid codes. The Fault ride-through (FRT) capability of WPPs is such a require-ment, the assessment of which requires a detailed dynamic simulation of the system, a simulation that encompasses numerous cyclic dependencies between different sys-tem components and the general maintenance of synchronism (i.e., transient and frequency stability). As a result, FRT serves as a rigorous test for co-simulation tools and simulation interfaces.

The main models involved in the implementation of the test case include dynamic models of the WPP, converter controller and FRT controller. The standard IEEE-9 bus system was modified to replace the generator at bus 3 by a WPP consisting of full converter interfaced generators. The WPP is connected to the rest of the grid at the point of common coupling (PCC). The PCC is significant since all the important metrics like compliance to appropriate voltage-time profiles during FRT is monitored and evaluated here and forms a legal boundary between the plant assets owner and the grid operator. The controllers developed are embedded in the AC-DC converter controller. A single-line diagram of the experimental setup can be seen in Fig. 5).

The WPP is an aggregated version of a medium to large scale onshore wind power park. The WPP is rated at 85 MVA which is cumulative rating of 32 wind turbines, each having a power rating of 2.6 MVA. The wind turbines are assumed to be deployed in an 8X4 array distanced by 700 m each.

The *converter controller* is designed with two proportional-integral controllers and one overarching current limiter. The controller is a *grid following* vector con-troller, which also models the reference voltage signals to control the voltages on both AC and DC side. The q-axis controller regulates the voltage magnitude of the PCC, whereas the d-axis controller maintains the active power reference [6].

The *FRT controller* acts on the top of the converter controller as a discrete finite state machine. It monitors the voltages on both AC and DC sides to sense fault conditions and shifts from normal control mode to FRT control, post-FRT mode,

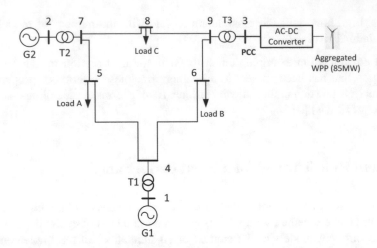

Fig. 5 Modified IEEE 9-bus system acting as a test system for co-simulating PowerFactory with Matlab/Simulink using the functional mock-up interface

and back to normal control accordingly. During the FRT mode, the FRT controller increases the reactive current infeed and blocks active power flow to address the voltage dip. In post-FRT mode, the FRT controller sets a maximum ramping rate for restoring the active current reference back to pre-fault conditions.

5.1 Experiment Setup and Objectives

The co-simulation is set up as follows. The AC grid including the wind part array is modelled and simulated in DIgSILENT PowerFactory. Inside PowerFactory, the wind turbine is modelled as a Norton equivalent source, the current injection of which can vary in time and is provided by the parameter event functionality as discussed in Sect. 3.1.1. This can be considered as a proxy model of the actual converter dynamics by the converter controller and FRT controllers, which are developed in Simulink and Matlab respectively. During runtime of the co-simulation, all FMUs are synchronised using fixed macro time step-sizes of 10 ms.

A 3-phase short circuit event is simulated to study the FRT capability. The co-simulation is orchestrated by a Python script, which uses the FMI++ toolbox. Eventually the overall system under test is split into three FMUs. The three FMUs being: One FMU for the entire power system model in PowerFactory, one for the FRT controller and one for the converter controller. The arrangement of the FMUs and the co-simulation orchestration by Python and FMI++ is shown in Fig. 6.

The simulation itself is centred around the dynamic response of the WPP and the IEEE 9-bus system that is subject to a self-cleared 180 ms 3-phase short circuit starting at $t = 1$ s (Bus 6 of Fig. 5). The main objective is to study the FRT capability

Fig. 6 Co-simulation Setup. On top the Python script using FMI++ to interface with the functional mockup units below. Left the FMU of PowerFactory based on FMI for co-simulation, in the centre the FMU of the converter (vector) controller based on FMI for model exchange. On the right the FMU of the fault ride-through controller based on FMI for model exchange. Both have an encapsulated dedicated numerical solver

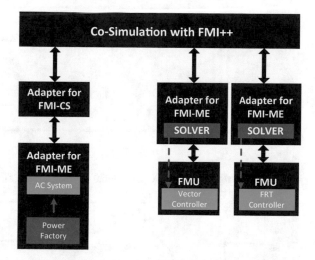

and reactive power control of the WPP. These ensure that the voltage at the PCC does not dip beyond the FRT voltage versus time profile, and quickly ramp the voltage to pre-fault levels after fault clearance (i.e., grid code compliance). During the event, the WPP shall remain connected to the grid. Adherence to these conditions during the simulations will certify that the co-simulation tools and interfaces have performed at their expected levels. In order to validate the co-simulation, a reference monolithic simulation was conducted in PowerFactory, too, in which standard dynamic models and dedicated DSL for the converter controls have been employed to duplicate the model specification in Matlab/Simulink.

5.2 Results

Figures 7 and 8 show the voltage magnitude at the PCC and active power through the PCC respectively. Taking the monolithic (PowerFactory only) simulation as a reference, it can be seen that the voltage sag experienced at the PCC is around 50% of nominal, which, quickly restores after fault clearance and swings back to values around nominal seconds after. This fast restoration is owing to the relatively strong grid as well as the voltage-dependent reactive current injection during the voltage dip. The presence of the active power recovery rate, engaged by the FRT controller, can also be clearly distinguished.

Despite the rather tight coupling between the submodels of the co-simulation, the dynamics around the PCC (red solid line) follow the reference simulation generally well. During fault ignition and clearance, a small discrepancy can be observed, particularly in the voltage magnitude, which can be attributed to the numerical oscillations caused by the serial data exchange protocol (see Fig. 4) that is applied in

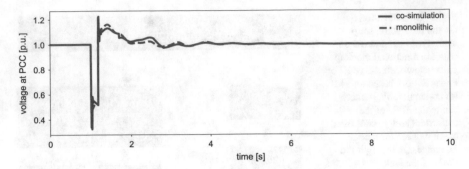

Fig. 7 Voltage magnitude at PCC, monolithic (red, solid) versus co-simulation (dashed, blue)

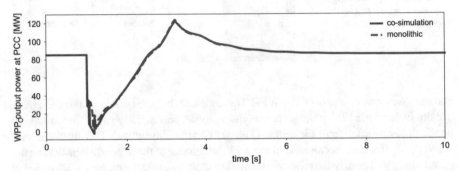

Fig. 8 WPP output active power, monolithic (red, solid) versus co-simulation (dashed, blue)

the master algorithm. Especially the active power, which can be considered a *flow* variable, traces the monolithic simulation very similarly. This enhances the validity of the co-simulation as a whole.

6 Conclusion

Major advancements in power electronic technology lead to its availability at all voltage levels and a massive deployment of components and systems grid-interfaced though power electronics. This incredibly boosts the controllability of the system as a whole but also introduces coupling of phenomena in the time domain that could normally be addressed separately such as power system stability. Likewise, the digital transformation drastically increased the heterogeneity of the electricity system, transforming it from a purely physical system, showing continuous behaviour, to a cyber-physical system also exhibiting discrete-event behaviour (non-linear, discontinuous).

Simulation bases assessment is a crucial link in the testing and validation chain of such integrated and intelligent energy systems (i.e. analysis, simulation, demonstration, roll-out). The heterogeneity of the (sub-)system models, however, shall be captured in the associated simulation tools accordingly. This is challenging as most simulators have been numerically optimised for a well-bounded domain, sometimes over decades. A solution for this challenge has been detailed in this chapter: co-simulation.

Various aspects of (coupling) simulation have been discussed: an overall typology for simulation-based assessment of CPES, the basics of co-simulation, standardised master algorithms and interfaces, and the framework approach adopted in the ERI-Grid project. More specifically, the ERIGrid project achieved the following additions to the state-of-art in terms of co-simulation

- Readiness of the mosaik co-simulation framework for mutually coupled subsystems in the time-domain (i.e., cyclic dependencies);
- Implementation of an FMI++ adapter in PowerFactory (RMS mode) complying with the FMI for co-simulation specification. An application example has been discussed in Sect. 5;
- Development of an FMI++ export package for ns-3 based on FMI for co-simulation;
- Implementation of the FMI++ MATLAB toolbox based on FMI for co-simulation; and
- Proof-of-concept of continuous-time, discrete-event, and mixed simulator coupling;
- Assessment of the scalability of the applied approaches; and
- Application of the holistic testing methodology for simulation-based assessment methods.

Notwithstanding these innovations in terms of applications of co-simulations, the approach is not very suitable for the day-to-day engineer yet. Parameterisation of simulator interfaces, master algorithm configuration, distributed execution, and harmonisation of semantics of the overall simulation are examples that require a lot of manual work and are still subject to technological development. Once this is mature, the benefits are unprecedented: simulation-based assessment of heterogeneous CPES as a service.

References

1. IEEE standard for modeling and simulation (m & s) high level architecture (hla)—object model template (omt) specification (2010)
2. Barenblatt, G.I.: Scaling. Cambridge University Press, Cambridge (2003)
3. Bhandia, R., van der Meer, A.A., Widl, E., Strasser, T.I., et al.: D-JRA2.3 Smart Grid Simulation Environment. Deliverable D8.3, ERIGrid Consortium (2018)
4. Blochwitz, T., Otter, M., Akesson, J., Arnold, M., Clauss, C., et al.: Functional mockup interface 2.0: the standard for tool independent exchange of simulation models. In: Proceedings of the 9th International MODELICA Conference, September 3–5, 2012, Munich, Germany, 076, pp. 173–184. Linköping University Electronic Press (2012)

5. Garro, A., Falcone, A.: On the integration of HLA and FMI for supporting interoperability and reusability in distributed simulation. In: Proceedings of the Symposium on Theory of Modeling and Simulation: DEVS Integrative M&S Symposium, pp. 9–16. Society for Computer Simulation International (2015)
6. van der Meer, A.A., Bhandia, R., Widl, E., Heussen, K., Steinbrink, C., Chodura, P., Strasser, T.I., Palensky, P.: Towards scalable FMI-based co-simulation of wind energy systems using powerfactory. In: Proceedings of Innovative Smart Grid Technologies (ISGT) Europe. Bucharest, Romania (2019)
7. Palensky, P., van der Meer, A.A., López, C.D., Jozeph, A., Pan, K.: Applied co-simulation of intelligent power systems: implementation, usage, examples. IEEE Indust. Electron. Mag. **11**(2) (2017)
8. Rohjans, S., Lehnhoff, S., Schütte, S., Scherfke, S., Hussain, S.: Mosaik - a modular platform for the evaluation of agent-based smart grid control. In: IEEE PES ISGT Europe 2013, pp. 1–5 (2013)
9. Steinbrink, C., van der Meer, A.A., Cvetkovic, M., Babazadeh, D., Rohjans, S., Palensky, P., Lehnhoff, S.: Smart grid co-simulation with MOSAIK and HLA: a comparison study. Comput. Sci. - Res. Devel. **23** (2017)

Hardware-in-the-Loop Assessment Methods

V. H. Nguyen, Q. T. Tran, E. Guillo-Sansano, P. Kotsampopoulos,
C. Gavriluta, G. Lauss, T. I. Strasser⬤, T. V. Jensen, K. Heussen, O. Gehrke,
Y. Besanger, T. L. Nguyen, M. H. Syed, R. Brandl, and G. Arnold

1 Introduction

The classical validation workflow in cyber-physical energy system (CPES) assessment is based on mainly two approaches: simulation and real-hardware testing. Simulation provides the advantages of rapidity, flexibility and versatility with no risk to damaging the equipment. Real hardware testing often requires more time and investments and is hard to reconfigure in case of necessity of adaptation, but it allows the consideration of real behaviour and impact of equipment that is usually hard to fully capture in a simulation environment.

Combining the strength of both approaches, advanced validation techniques interfacing real and virtual environment such as: real-time (RT) simulation [2], controller-hardware-in-the-loop (CHIL) and power-hardware-in-the-loop (PHIL) [5] or eventu-

V. H. Nguyen (✉) · Q. T. Tran
Université Grenoble Alpes, INES, CEA, LITEN,
Le Bourget du Lac, France
e-mail: vanhoa.nguyen@cea.fr

E. Guillo-Sansano · M. H. Syed
University of Strathclyde, Glasgow, Scotland, UK

P. Kotsampopoulos
National Technical University of Athens, Athens, Greece

C. Gavriluta · G. Lauss · T. I. Strasser
AIT Austrian Institute for Technology, Vienna, Austria

T. V. Jensen · K. Heussen · O. Gehrke
Technical University of Denmark, Roskilde, Denmark

Y. Besanger · T. L. Nguyen
Université Grenoble Alpes, Grenoble INP, Grenoble, France

R. Brandl · G. Arnold
Fraunhofer Institute for Energy Economics and Energy System Technology IEE,
Kassel, Germany

© The Author(s) 2020
T. I. Strasser et al. (eds.), *European Guide to Power System Testing*,
https://doi.org/10.1007/978-3-030-42274-5_4

ally the combination of both techniques [8] are new and efficient testing methods for Distributed Energy Resource (DER) devices, for manufacturers to adapt their products to the increasingly demanding requirements, as well as for network operators and regulation authorities to establish new testing and certification procedures on a system point of view. In these advanced Hardware-in-the-loop (HIL) techniques, a real hardware setup for a domain (or part of a domain) is coupled with a real-time simulator to allow testing of hardware or software components under realistic conditions. HIL provides the advantage of replacing error-prone or incomplete models with real-world counterparts and the possibility of scalable testing in faulty and extreme conditions. Real-Time Simulation (RTS) and HIL have proven their applicability as the upcoming and future methodology for testing the (future) smart grid, including DER devices and ICT network to form a holistic and modern power system.

In this chapter, HIL methods are considered as potential methods for configuring complex and realistic validation environment for smart grid. We present the necessary considerations in setting up a HIL experiment, i.e. stability assessment and latency compensation and we propose several approaches for the integration of HIL techniques to a holistic testing framework. This chapter can introduce some insights and technical solutions to readers to create more sophisticated and more realistic experiments.

2 HIL Techniques for Validation of Smart Grid Solutions

The usage of HIL techniques in smart grid applications is generally classified into CHIL and PHIL [1, 2]. A general HIL setup consists of three main elements, the RT simulator, the HUT, and the power interface (only in PHIL case) as depicted in Fig. 1. The RT simulator computes the simulation model in real-time and offers Input/Output (I/O) interfaces/channels to reproduce the behaviour of the simulated system under dynamic conditions. The simulator allows designing and performing various test scenarios with a great flexibility.

Controller Hardware-in-the-Loop (CHIL) involves the testing of a device, for example a power converter controller, where signals are exchanged between a Real-Time (RT) simulator and the HUT via its information ports. The interface in that case (CHIL) consists of Analogue to Digital and Digital to Analogue converters

Fig. 1 Basic elements of a (P)HIL experiment

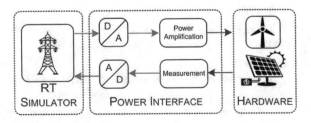

and/or digital communication interfaces. Besides control devices as although real-time simulations coupled to other units such as relays, Phasor Measurement Units (PMU) or monitoring devices are usually classified as CHIL. Such devices are validated in a closed-loop environment under different dynamic and fault conditions, therefore enhancing the validation of control and protection systems for power systems and energy components. In contrast, PHIL involves the testing of a device which absorbs or generates electrical power (e.g., Photovoltaic inverter). A power interface is therefore necessary. CHIL allows testing of physical controller devices, such as DER controllers, relays, PMU, etc., while PHIL involves also a wide variety of DER devices and network components such as converter, electric vehicles and corresponding charging equipment (Fig. 1).

Extending the concept of PHIL, also whole micro-grids or distribution grids can be tested in realistic environments. To distinguish the interfacing of single hardware components from the coupling of multi-device power hardware, the term Power System in-the-Loop is introduced (PSIL). In a broader frame, HIL testing can thus be extended to laboratories offering full-scale physical setups in which pure hardware interactions of multiple components and distributed control hardware become part of the experiment. In this sense PSIL challenges the sharp distinction between CHIL and PHIL and offers a future perspective of hybrid experiments of power hardware, a power network configuration, and control hardware/software.

Along with the increased realism of experiment, one major challenge of going from pure simulation to PSIL is the high complexity of implementing real-time compliant interfaces between the different elements/domains (e.g. RTS, controller, electrical components, SCADA system). The issues such as communication latency and stability of the interface must be assessed properly to avoid damage to the physical equipment and to the users.

2.1 Stability of HIL Experiments

Due to the addition of the hardware/simulation interface, thus various external disturbances (especially the time delay), HIL experiments are sensitive in terms of stability and accuracy. Additionally, for PHIL, the power amplification configuration and its impact (I/O boundaries, galvanic isolation, short circuit behaviour, slew-rate, etc.) must be addressed and evaluated as it strongly influences the determination of system stability, bandwidth, and the expected accuracy. Instability in PHIL simulations should be avoided as it can cause irreparable damage to equipment.

2.2 Stability Assessment

The model of a PHIL simulation can be expressed using transfer function in the frequency domain as shown in Fig. 2.

Fig. 2 General
representation of a PHIL
system

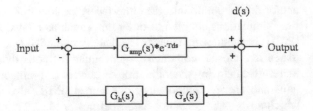

$G_s(s)$, $G_{amp}(s)$ and $G_h(s)$ are the transfer functions of the simulated part, amplifier and hardware part respectively and the exponential term is the representation in
the frequency domain of the time delay inserted by the amplifier. The disturbance
inserted into the system due to extrinsic factors is noted as d(s).

Using Bode stability criterion the stability conditions can be expressed as:

$$|G_s(s)G_{amp}(s)e^{sT_d}G_h(s)| < 1 \text{ and } \angle G_s(s) + \angle G_{amp}(s) + \angle G_h(s) - \omega T_d = \pi$$

Taking into account the uncertainties that occur in different parts of the model of a
PHIL simulation the previous inequality related with magnitude of the open transfer
function is given by:

$$|G_s(s)G_{amp}(s)e^{sT_d}G_h(s)| < \frac{1}{1+\epsilon}$$

As the parameter ϵ is, by definition, always bigger than zero, the value of the fraction
of the right part of the inequality is smaller than unity. Thus, one can conclude that
when there are unmodeled parts in the system intentionally or not, then the stability
criterion of the analysis should be stricter. From a practical point of view, we apply
a more conservative method in order to examine the bounds of the stability of the
system and to derive safe results even in the worst-case scenario. Moreover, based on
the Bode stability criterion the marginal parameters of a PHIL experiment (to achieve
stability) can be determined without using approximations for the time delay. The
proposed analysis has been applied to existing methods to achieve stability in [7].

2.3 Approaches for the Compensation of Time Delay

The time delay presented in HIL simulations directly affects the phase relationship
of the signals exchanged at the point of common coupling and accordingly the power
factor of the HUT seen by the simulation platform [3, 6, 14]. Furthermore, this effect
will not only be present at the fundamental frequency but also at any harmonic component present in the simulation, hence the time delay in these harmonic components
should also be reduced or compensated.

The action of compensating the time delay aims to achieving a waveform which
is in phase and has the same amplitude as that of an ideal system (without PHIL
interface). With this purpose, different approaches exist which by the application of
a filter or a phase-shift results in a non-delayed waveform. When a phase-shift is

applied, this needs to be applied to all the harmonic components of the waveform for an accurate solution. The main approaches used for the compensation of time delay are:

- Fourier compensation: by applying Fourier transformation to the signals, the phasors of the harmonic components can be identified. Then, the time delay compensation can be performed by leading the phase for each harmonic of interest [3].
- Lead compensator: this function can be used for approximating the interface to an ideal interface in terms of phase and gain [12, 13]. This approach is susceptible of amplifying high frequency noise and can be limited when harmonic components are present or the system under test is complex.
- Direct-quadrature-zero (dq0) transformation: different harmonic components can be identified with the dq0 transformation, which can be independently compensated with the addition of the time delay to the inverse dq0 transformation [6]. However, this approach relies in noise free and balanced three phase waveforms and can be computationally expensive if large number of harmonics are required.
- Synchronous generator control compensation: phase compensation can be introduced in the control algorithm of the synchronous generator when it is used as the power amplifier for PHIL [14].

Complimentary to the time delay compensation, to achieve improved transient dynamics in PHIL simulations the total time delay has to be minimized. This can be achieved by (i) minimizing the time step of the real time simulation, (ii) reducing time delays introduced by the power amplifiers, (iii) improving the communication channels used (with fast digital communications rather than analogue communication), and (iv) avoiding the use of components that add significant delays such as filters.

3 Integration of HIL Techniques into a Holistic Framework

By the date of publication of this book and to the extent of the authors knowledge, there are no off-the-shelf tools available for testing complex smart grid applications that involve components from different domains. In order to achieve a complete validation of the multi-domain and large-scale smart grid, HIL techniques can be combined with other simulations and with other infrastructures. The idea of integrating real-time based HIL to a holistic framework provides the basis of the subsequent ERIGrid's approaches. With these solutions, an integral view of the behaviour of the communication network and the states of the power system may be achieved.

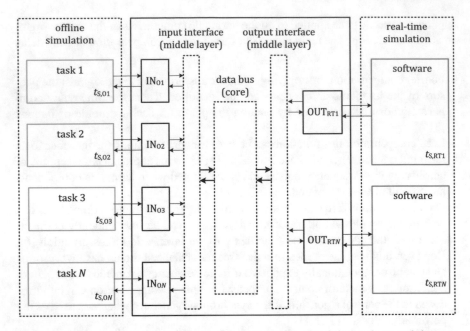

Fig. 3 Simulation message bus architecture for co-simulation of real-time and non-real-time systems

3.1 Simulation Message-Bus Based Solutions: Lab-Link and OPSIM

The principal architecture of an Simulation Message Bus (SMB)-based co-simulation is presented in Fig. 3. In this configuration, the main component is represented by the simulation data bus. Input and output interfaces are packed around the core and act as a middle layer allowing data structures to be injected or extracted from the message bus. Depending on the sample rate at which data needs to be exchanged with the core, specifically designed task processing units will be needed for the purpose of the respective application.

These task processing units represent functional units which are typically implemented in software and are labelled as IN_{O1}, IN_{O2}, IN_{O3}, ... IN_{ON} in Fig. 3. Their primary function is to design custom software or hardware adaptations for each application or simulator that participates. The SMB may be modified in the course of the development of various co-simulation tools.

Lablink is a software package based on SMB, as indicated before. The purposely designed communication middleware allows for fast and simple coupling of software and hardware components. Mainly, Lablink enables different devices integrated in an electric power laboratory such as power sources, loads, grid emulators, or measurement devices to establish a bidirectional data and control signal flow. Figure 4 shows the basic structure of the Lablink applicable for real-time and non real-time

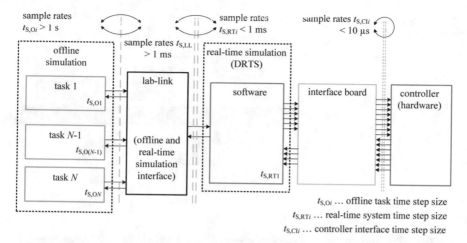

Fig. 4 Lablink structure for real-time and non real-time CHIL applications with indicated sample rates

simulation. The left part highlights N offline simulation tasks with time step sizes assumed in the ranges of $t_{S,Oi} \in [100 \text{ ms}; 2 \text{ s}]$. All mentioned tasks are connected to the Lablink structure in an independent and bidirectional way with respect to signal or data exchange. The range of the time step sizes may heavily vary based on the type of offline simulation. However, typical values are proposed in Fig. 4 for simulation setups related to investigations in the electrical domain.

As highlighted in the SMB architecture shown in Fig. 4, Lablink is processing incoming and outgoing data from offline simulation tasks and from the linked DRTS, respectively. In this case, minimum time step sizes of $t_{S,LL} = 1$ ms are specified as sample rates for Lablink. However, the real-time computing system has fixed time step sizes due to the inherent constraint of real-time simulation. For CHIL applications, the DRTS typically runs with a time step size in the range of $t_{S,RTi} \in [100 \text{ ns}; 1 \text{ ms}]$. Sample rates of less than 1 µs are required for simulation tasks emulating PWM signals for control application.

As shown in Fig. 4, the real-time machine on the right side is linked to one or several interface boards. The interfacing boards represent functional units between machine controller implemented in hardware and the DRTS system. The number of signals exchanged between the controller and the DRTS may be high. At least, it is higher than the number of signals exchanged between offline tasks and Lablink for typical CHIL applications related to converter control simulations. The maximum specified time step size $t_{S,CIi}$ referring to the controller interface is set by 10 µs.

OpSim (Fig. 5), another SMB-based solution, enables users to connect their software to simulated power systems or test it in conjunction with other software. The core of OpSim is a flexible message bus architecture; it allows arbitrary co-simulations in which power system simulators, controllers and operative control software can be coupled together. The OpSim Message Bus works as a unidirectional buffer with validity during a time window defined by the publish rate T_x of the simulator which

Fig. 5 OpSim solution architecture

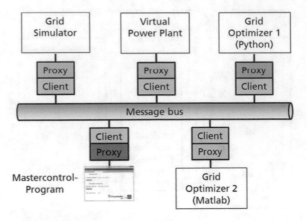

Fig. 6 OpSim Message Bus data handling with two simulators

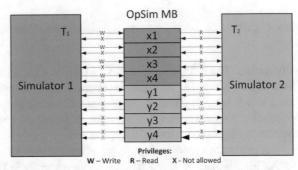

has write privileges on the variable/vector to exchange. This time window prevents corruption or overwriting of the information from external sources and ensures a real-time simulation according to each simulator publish rate. This leads to defining a variable for each simulator-to-simulator value exchange, as can be seen in the Fig. 6.

3.2 Online Integration with SCADA as a Service Approach

In this approach, a hybrid cloud server (Fig. 7) is used as the intermediate buffer for information exchange among elements of a holistic test. In order to improve interoperability and reusability of the developed models, the Functional Mock-up Interface (FMI) standard can be integrated. This approach allows the integration of RTS to multiple hardware (SCADA, DER, etc.) and software (simulators). The synchronization is configured to satisfy the conditions in [10].

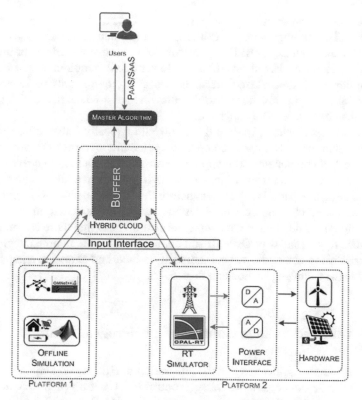

Fig. 7 Multi-infrastructure integration via SCADA as a service approach [11]

3.3 Quasi-static PHIL/PSIL

The Quasi-static HIL approach addresses applications in the smart grid context that do not require very fast analysis, such as optimization evaluation or use of secondary control. In these applications, the requirements of hardware and associated simulators need not comply with strict real-time requirements. As an example, for investigating voltage support controllers in distribution grids, this variant of the PSIL approach could couple a real power system, e.g. a LV feeder, to a simulated grid, e.g. MV grid with multiple feeders, and associated controllers. Since the objective of the experiment does not directly address fast phenomena, the investigation can proceed with a relaxed real-time constraint and the integration of slower, non-real-time, simulations into HIL. Under assumption of the relaxed testing requirements outlined in [10] and the extension of supported domain couplings and software interfaces, there is thus room for a relaxed variant of hardware-software coupling in laboratory experiments, which we summarize under the term "Quasi-static Power Hardware In-the-Loop" (QsPHIL) [9].

Testing smart grid solutions under QsPHIL should be seen as complementary to PHIL and CHIL testing: (a) since QsPHIL assumes that the electrical grid can be treated as quasi-stationary, QsPHIL implicitly assumes that the effects of transients and stability can be neglected, and (b) due to lower fidelity requirements of simulator and coupling hardware, relaxed precision and cost requirements make PSIL and multi-domain experiments more flexible, allowing for a wider range and scope of experimental hardware to be integrated.

Thus, in a testing chain, QsPHIL testing would typically follow after PHIL and CHIL tests of individual hardware components or assume maturity of power hardware and fast control components. The approach can also be assumed to cover field test implementation in a controlled laboratory environment, making it suitable as an intermediate step or partial replacement before field deployment. This is particularly true for validation of complex control systems, where implementation errors can be caught before field deployment, reducing the cost of managing these errors. Further, the reduced requirements of QsPHIL suit applications for the remote integration of laboratories using non-dedicated communication links and hardware equipment, as reported in the following section and Sect. 5.4.5.

4 Coordinated Voltage Control of a Microgrid Example

To demonstrate the application and the integration of HIL to a co-simulation framework, a test-case of coordinated voltage control (CVC) of a benchmark microgrid [4] is implemented in CHIL manner (via (Lab-link and OpSim architecture) and then in PHIL-PSIL manner (via SCADA-as-a-service approach). The benchmark microgrid (modified from the CIGRE LV grid) is governed by a CVC algorithm aiming to minimize bus voltage deviation, power loss and the number of tap change by the OLTC (Fig. 8). The objective of the test-case is validating the performance of the CVC algorithm and demonstrating the implementation of a complex and realistic validation environment using HIL techniques.

4.1 CHIL Implementation via Lab-Link

The first demonstration is implemented in a single infrastructure (AIT Austrian Institute of Technology—Austria). The lab-link architecture is used to implement the test-case, linking the microgrid model in real-time simulation (by OPAL RT), and the controller (in Matlab) (Fig. 9).

The bus voltage in the microgrid is then regulated according to the three optimization criteria (Fig. 10).

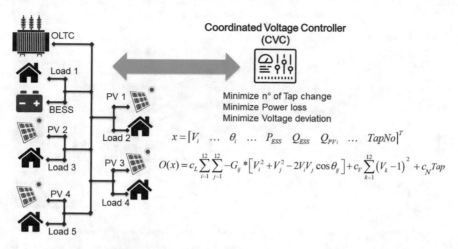

Fig. 8 The benchmark microgrid and the CVC algorithm

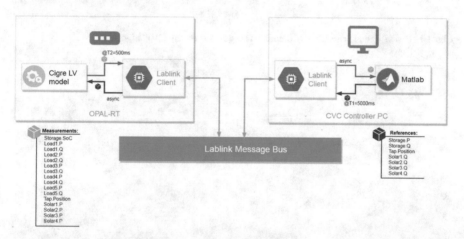

Fig. 9 Implementation of the test-case with Lab-link architecture

4.2 Multi-platform CHIL Implementation via OpSim Architecture

Using the OpSim environment, the test-case is implemented in a multi-infrastructure manner, with the microgrid simulated by OPAL RT and connected directly to the OpSim message bus at Fraunhofer IEE in Kassel—Germany and the controller running in Matlab at National Technical University of Athens—Greece, connected to the co-simulation environment via the OpSim web interface (Fig. 11).

The result in Fig. 11 shows slight deviations of voltage with respect to the single platform implementation, demonstrating (i) the combination of expertise and equip-

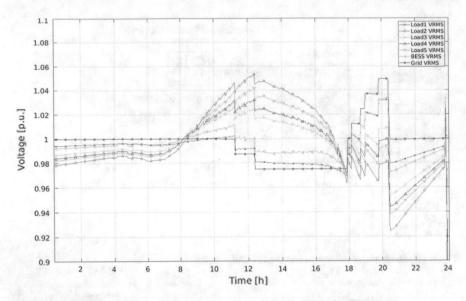

Fig. 10 Bus voltage at different grid nodes as regulated by the CVC algorithm

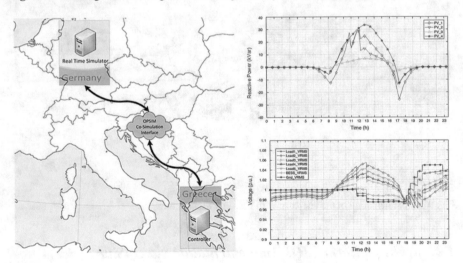

Fig. 11 Cross-Infrastructure implementation with OPSim and results (PV Reactive Power and Bus voltages

ment of the two infrastructures in a holistic test and allowing (ii) the consideration of latency's impact to the performance of the control. In this case, the CVC algorithm still shows good performance despite of the influence of latency between the two platforms (Fig. 11).

Fig. 12 The houses and PV packs considered in the microgrid

4.3   PHIL and PSIL Implementation in PRISMES Platform

The behaviour of the proposed CVC algorithm is then tested in a more realistic environment, a real microgrid in PRISMES platform (Commissariat of Atomic Energy and Alternative Energies—CEA France). The experiment is implemented with a PHIL part interfacing a physical load and a PV inverter (PV 4 and Load 5) with the real-time simulator OPAL RT mimicking the real microgrid behaviours via its SCADA system (SCADA as-a-service approach). The setup can be considered as an approach of Power-System-In-The-Loop (PSIL). In this case, the loads 1–4 (Fig. 8) are replaced with the digital twins of 4 experimental smart houses INCAS and the PV pack from 1 to 3 are replaced with the digital twins of three real PV packs (20-20-60 monocristalin panels and inverters) in PRISMES platform (Fig. 12).

To this purpose, the digital twins of the equipment are replicated in real-time simulation with the OP5700 RT simulator with their measurements synchronized from the SCADA System. The synchronization step for each measure is chosen according to the conditions proposed in [10] and with respect to: 1/ Physical sampling time of the physical sensors, 2/ Latency between the SCADA server and the RT Simulator. The RT simulator are also responsible for simulating the equipment that are not physically available in the platform (i.e. OLTC and BESS). Moreover, to facilitate the study on impact of the CVC algorithm on radial ends of the microgrid, load 5 and PV 4 are replaced with real equipment (1 PV pack with SMA inverter and 1 load Cinegia) (Fig. 13) and are connected to the grid via the PHIL interface (Fig. 14). The inclusion of PHIL part also allows consideration of more advanced functionalities of the integrated hardware (e.g. Fault-ride-through or anti-islanding).

The proposed PSIL setup presents several advantages and is much more realistic and challenging for testing the CVC for the following reasons:

- The combination of simulation and real equipment provides great flexibility in configuring complex, yet realistic validation environments.

Fig. 13 The PV inverter and the load in PHIL integration

Fig. 14 The RT simulator OP5700 and the Puissance+ Power Amplifier used in the PHIL interface

- The PV production and loads reflect exactly a real deployment environment (irradiation angle, weather condition, load demand, etc.)
- The real-time digital twins is synchronized with real measurement from the SCADA system, which is much more intermittent and is subjected to a wide range of disturbance from the communication network and the SCADA service itself.
- All the INCAS houses are energy positive (equipped with rooftop PV). So the microgrid presents a very high PV penetration rate.

The CVC algorithm is then applied to this PSIL setup to control the microgrid voltage according to the desired criteria (Fig. 15). OLTC and BESS (virtual equipment) are regulated together with the reactive power of PV inverter (real equipment in PHIL and digital twins of real equipment) to act on the bus voltages, deviated by the PV production and load. The impact of PV injection can be studied as the voltage increases on radial ends (i.e. PV 4 and load 5).

The three selected test-cases represented different levels of validation of a CVC algorithm, from CHIL to CHIL with consideration of latency and finally PHIL and PSIL setup. They demonstrate the potential applications of RTS and HIL techniques in configuring complex and realistic validation environments for smart grid, according to user's needs.

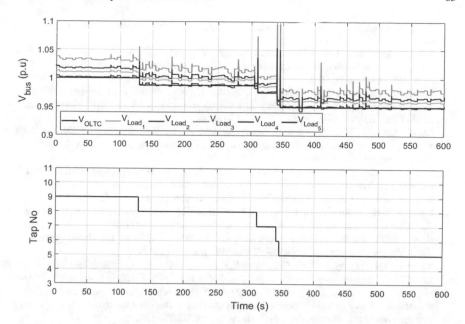

Fig. 15 The CVC algorithm regulates the microgrid voltage according to the desired criteria w.r.t real PV production and consummation

5 Summary

CPES constitute a significant challenge for system testing and validation. In this chapter, the development of HIL experiments to tackle the validation and assessment of holistic and complex smart grid scenarios were presented, along with its associated laboratory environment. These solutions, involving advanced techniques such as integration with co-simulation, stability assessment and time delay compensation, provide flexible and equally reliable testing environments for various smart grid setups with different degrees of complexity. They are demonstrated via three selected test-cases representing different implementation levels of hardware inclusion, from CHIL to PHIL and PSIL.

References

1. De Jong, E., De Graff, R., Crolla, P., Vassen, P., et al.: European White Book on Real-Time Power Hardware in the Loop Testing. Technical report, DERlab Consoritum (2012)
2. Faruque, M.D.O., Strasser, T., Lauss, G., Jalili-Marandi, V., et al.: Real-time simulation technologies for power systems design, testing, and analysis. IEEE Power Energy Technol. Syst. J. 2(2), 63–73 (2015)
3. Guillo-Sansano, E., Roscoe, A.J., Burt, G.M.: Harmonic-by-harmonic time delay compensation method for PHIL simulation of low impedance power systems. In: 2015 International

Symposium on Smart Electric Distribution Systems and Technologies (EDST), pp. 560–565 (2015)

4. Kotsampopoulos, P., Lagos, D., Hatziargyriou, N., Faruque, M.O., et al.: A benchmark system for hardware-in-the-loop testing of distributed energy resources. IEEE Power Energy Technol. Syst. J. **5**(3), 94–103 (2018)

5. Lauss, G.F., Faruque, M.O., Schoder, K., Dufour, C., Viehweider, A., Langston, J.: Characteristics and design of power hardware-in-the-loop simulations for electrical power systems. IEEE Trans. Indust. Electron. **63**(1), 406–417 (2016)

6. Li, G., Jiang, S., Xin, Y., Wang, Z., Wang, L., Wu, X., Li, X.: An improved DIM interface algorithm for the MMC-HVDC power hardware-in-the-loop simulation system. Int. J. Electr. Power Energy Syst. **99**, 69–78 (2018)

7. Markou, A., Kleftakis, V., Kotsampopoulos, P., Hatziargyriou, N.: Improving existing methods for stable and more accurate power hardware-in-the-loop experiments. In: 2017 IEEE 26th International Symposium on Industrial Electronics (ISIE), pp. 496–502 (2017)

8. Nguyen, V.H., Besanger, Y., Tran, Q.T., Boudinnet, C., Nguyen, T.L., Brandl, R., Strasser, T.I.: Using power-hardware-in-the-loop experiments together with co-simulation for the holistic validation of cyber-physical energy systems. In: 2017 IEEE PES Innovative Smart Grid Technologies Conference Europe (ISGT-Europe), pp. 1–6 (2017)

9. Nguyen, V.H., Bourry, F., Tran, Q.T., Brandl, R., et al.: D-JRA3.2 Extended Real-Time Simulation and Hardware-in-the-loop Possibilities. Technical report, ERIGrid Consortium (2018)

10. Nguyen, V.H., Nguyen, T.L., Tran, Q.T., Besanger, Y.: Synchronization conditions and real-time constraints in co-simulation and hardware-in-the-loop techniques for cyber-physical energy system assessment. Sustain. Energy, Grids Netw. **20**, 100252 (2019)

11. Nguyen, V.H., Tran, Q.T., Besanger, Y.: SCADA as a service approach for interoperability of micro-grid platforms. Sustain. Energy, Grids Netw. **8**, 26–36 (2016)

12. Pokharel, M., Ho, C.N.M.: Stability study of power hardware in the loop (PHIL)simulations with a real solar inverter. In: Proceedings IECON 2017 - 43rd Annual Conference of the IEEE Industrial Electronics Society, vol. 2017, pp. 2701–2706 (2017)

13. Ren, W., Sloderbeck, M., Steurer, M., Dinavahi, V., et al.: Interfacing issues in real-time digital simulators. IEEE Trans. Power Deliv. **26**(2), 1221–1230 (2011)

14. Roscoe, A.J., MacKay, A., Burt, G.M., McDonald, J.R.: Architecture of a network-in-the-loop environment for characterizing AC power-system behavior. IEEE Trans. Indust. Electron. **57**(4), 1245–1253 (2010)

Laboratory Coupling Approach

L. Pellegrino, D. Pala, E. Bionda, V. S. Rajkumar, R. Bhandia, M. H. Syed,
E. Guillo-Sansano, J. Jimeno, J. Merino, D. Lagos, M. Maniatopoulos,
P. Kotsampopoulos, N. Akroud, O. Gehrke, K. Heussen, Q. T. Tran,
and V. H. Nguyen

1 Introduction

Nowadays there are several Research Infrastructures in Europe performing experiments on Smart Grid activities. Each of them has a particular strong point: hardware, software, models, procedures or the specific experience of the researchers. In order to exploit the potential of each one and to make it available to the community, the ERIGrid project developed advanced system validation methods and tools, together with common models, harmonized validation and deployment procedures.

L. Pellegrino (✉) · E. Bionda
Ricerca Sistema Energetico, Milan, Italy
e-mail: luigi.pellegrino@rse-web.it

D. Pala
A2A, Milan, Italy

V. S. Rajkumar · R. Bhandia
Delft University of Technology, Delft, The Netherlands

M. H. Syed · E. Guillo-Sansano
University of Strathclyde, Glasgow, Scotland, UK

J. Jimeno · J. Merino
TECNALIA Research & Innovation, Derio, Spain

D. Lagos · M. Maniatopoulos · P. Kotsampopoulos
National Technical University of Athens, Athens, Greece

N. Akroud
Ormazabal Corporate Technology, Boroa, Spain

O. Gehrke · K. Heussen
Technical University of Denmark, Roskilde, Denmark

Q. T. Tran · V. H. Nguyen
Université Grenoble Alpes, INES, Le Bourget du Lac, France

CEA, LITEN, Le Bourget du Lac, France

© The Author(s) 2020
T. I. Strasser et al. (eds.), *European Guide to Power System Testing*,
https://doi.org/10.1007/978-3-030-42274-5_5

One of the approaches developed in ERIGrid is the Laboratory coupling which allows to exploit the synergies among the research infrastructures. By sharing the hardware and software devices of different research infrastructures, new complex test cases can be performed on an extended system configuration: this favours technology development and facilitates the deployment phase so that the necessary engineering efforts are reduced and the time to market of innovations and solutions is shortened.

This section explains, first, the state of art for smart grids testing, and discusses the procedures and system configurations normally adopted for components or systems testing. Then, after some aspects related to the interoperability are introduced, the communication infrastructure developed in the framework of ERIGrid which allows to exchange data among the research infrastructures is discussed. Finally, Sect. 3 shows the new testing approaches enabled by coupling different RIs whereas in Sect. 4 four implementations of laboratory coupling are presented:

- integration of a remote OLTC controller via IEC 61850;
- state estimator web service;
- hardware/software integration between different research infrastructures;
- integration of remote hardware among different research infrastructures.

1.1 State-of-the-Art for Smart Grid Testing

The state of the art for smart grid testing is still lacking and in some cases is unclear. This is due to the high number of possible system configurations, functionalities and technologies to test in a smart grid environment. Some standardization bodies have already been developed together with specific testing procedures on particular aspects of smart grids. For instance, IEEE has published a standard for the interconnection and interoperability between the electric power systems and DERs [1]. This standard includes requirements, response to abnormal conditions, power quality, islanding, and test specifications and requirements for design, production, installation evaluation, commissioning, and periodic tests. Another standard, at the moment inactive, provided by IEEE, is related to the testing of microgrid controllers [2]. This standard wants to provide testing procedures which allow to test the energy management system of microgrids ensuring the "plug and play" functionality and establishing comparative performance indices.

Many other standards have been published and reviewed by national and international standardization bodies but there are still many testing procedures not included in these standards. This is the reason that brings research centres to develop and implement custom testing procedures. Focusing on the system configuration of a smart grid, four main testing approaches can be applied:

- *Simulation:* this testing approach allows to simulate the behaviour of the system configuration based on its mathematical model. Typically, this is the first experiment performed since it demonstrates the functionality of the technology under development. However, since the system configuration is only a mathematical model, many characteristics of the real system under test might be neglected. This could affect the results of the testing, hence further experiments, with increased reality, have to be performed.

- *Hardware In the Loop:* as a second round of testing, a HIL experiment might be performed. This technique, as deeply described in Sect. 4, allows to test hardware or software components under realistic conditions, coupling a real hardware setup for a domain (or part of a domain) with a real-time simulator. In this case the system under test includes real components, hence the test validation is very close to the field testing; only the part of the system under test simulated in the real-time simulator is a model.
- *Pure Hardware:* similar to the field testing, an experiment can also be performed in a pure hardware system configuration. In this case, the whole system configuration is composed of real devices; the behaviour of the system is exactly the real one in case of field testing or it is very similar in case of laboratory testing.
- *Combination of different testing approaches:* the system configuration of some tests requires, on one hand, a high reality level and, on the other hand, an extended system under test. These needs could not be satisfied by only one of the previous testing approaches. However, the combination of two or more testing approaches can enable some of these tests. This approach is beyond the state of the art, even considering the combination of different testing approaches in only research infrastructure. Indeed, the integration of different systems introduces several challenges. The problem is even worse in case of integration of multiple research infrastructures.

1.2 Multi-infrastructure Integration

Due to the increasing complexity of smart grids, an integrated approach for analysis and evaluation requires a large-scale validation scenario and may be unfeasible in one single research infrastructure (RI). Reflecting the interdisciplinary and dynamic nature of the field of smart grid research, many smart grid laboratories are designed to support a broad range of testing activities, from component testing to system testing, from hardware to software tests, from research to certification and education. This increased demand for flexibility is hard to achieve without also increasing the complexity of the laboratory infrastructure. Firstly, a combined expertise of different domains is required, which is not always the case of current specialized laboratory system; and secondly, the required complete RI for large scale CPES is theoretically possible but realistically not reasonable solution, in term of investments (equipment and expertise), operation (staff) and organization. Establishing an RI coupling framework allows the creation of a common resources and expertise pool to efficiently use the existing equipment and combine it with the complementary counterpart from others to validate researches in a holistic and near real-world environment.

Fig. 1 Interoperability
architecture in
cross-infrastructure holistic
experiment [3]

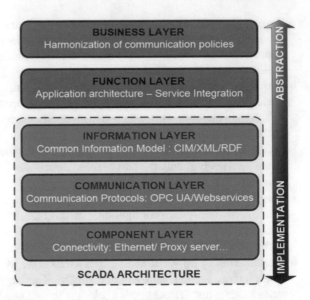

On the other hand, developing such a holistic validation framework for CPES would also benefit researchers in terms of facilitating the replication of experiments and the verification of the validity of the results.

The technical obstacles for laboratory collaboration are more narrowly related to the interoperability among these infrastructures. In [3], a generic five layers of interoperability among a consortium is proposed (Fig. 1). The top layer involves the harmonization and agreement on information sharing policies (i.e. legal and admin support). The conceptual and semantic design of the holistic test, derived from the desired scenario and the individual RI's capabilities, requires functional layer interoperability. The technical integration of RIs (i.e. actual interconnection) is deployed thanks to the three lower layers interoperability with possible involvement of SCADA architecture. Harmonization of information models (e.g. CIM), communication protocols as well as the aspect of synchronization, handling causality and latency compensation are therefore required for the seamless communication among infrastructures, the good emission and reception, and the correct interpretation of received information. This task is however not trivial due to the lack of flexible information models covering both power and ICT domains and due to the lack of efforts to harmonize the excessive numbers of communication protocols in the literature.

Two important aspects of interoperability in a laboratory context are the exchange of information between technical devices, and the deployment and/or execution of test-specific third-party software on the infrastructure. In the case of external access to a laboratory, necessary software adaptations caused by deployment constraints or lack of a suitable Application Programming Interface (API) may be a major part of the effort of integrating the laboratory. In extreme cases, software may have to be rewritten to adapt to the target environment. Another potential obstacle is related to the differences in security and confidentiality policies between research infrastructures. Figure 2 describes different interfacing possibilities for integrating an external (third party) element (equipment, controller, etc.) to the local infrastructure:

E1 Direct communication between a laboratory internal supervisory controller and a third-party controller, for example to allow the third party controller to influence the control behaviour of the supervisory controller.

E2 Direct communication between a laboratory internal supervisory controller and a third party SCADA system, for example to allow test sequencing software to control a third-party test device which is bringing its own SCADA system to the test.

E3 Direct communication between a third-party controller and the laboratory internal SCADA system, for example to allow the third party controller to control a laboratory internal DER unit.

E4 Direct communication between the laboratory internal SCADA system and a third party SCADA system, for example to integrate equipment controlled by the third party SCADA system into the laboratory SCADA system in order to control all equipment through a single interface.

E5 Direct communication between the laboratory internal SCADA system and a third party IED, for example to allow test software to control both lab components and external test components through a single SCADA interface.

E6 Direct communication between a third party SCADA system and a laboratory internal IED, for example to integrate a laboratory device into a test setup consisting of third party devices which are controlled by a third party SCADA system.

E7 Direct communication between a third party IED and a laboratory internal IED, for example in order to allow an external IED to influence the behaviour of the internal IED.

E8 Direct communication between a laboratory internal IED and an external DER controller, for example to control a third-party device from the laboratory SCADA system through a laboratory RTU.

E9 Direct communication between a third party IED and a laboratory internal DER controller, for example to test a third party IED (e.g. a site controller) against a laboratory DER unit.

E10 Direct communication between a third party DER controller and a laboratory internal DER controller, for example in order to enable load sharing between multiple generator sets.

E11 Direct communication between a laboratory internal DER controller and a third party DER unit, for example to control a compatible third-party device from the laboratory SCADA system through the lab DER controller.

E12 Direct communication between a third party DER controller and a laboratory internal DER unit, for example in order to test a third party DER controller against a laboratory DER unit.

In general, for each RI coupling interface, it is required to satisfy at least the three lower interoperability layers as described in Fig. 1. In terms of supported interface

Fig. 2 Different possibilities
for multi-infrastructure
integration

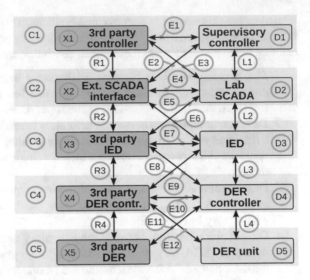

protocols, a wide variety of solutions is found including TCP/IP, UDP/IP (ASN.1 encoding according to IEC 61499), Modbus/Sunspec, IEC 61850, Java RMI, Matlab API, XML-RPC, OPC, proprietary interfaces and many more.

2   JaNDER Communication Platform for Lab-Coupling

This section provides an overview of the communication platform developed in the ERIGrid project which allows to exchange data online. This could be used for coupling different RIs: testing a software or a controller in a remote laboratory, acquiring data from several RIs or even creating a virtual research infrastructure. This platform is called "Joint Test Facility for Smart Energy Networks with Distributed Energy Resources" (JaNDER). JaNDER is a cloud platform for the exchange of information (measurements, control signals, laboratory asset descriptions) between geographically distributed labs by using a secure internet connection. This section describes the three different JaNDER levels developed.

2.1   Features of the Cloud-Based Communication Platform

Based on the needs of the possible users of JaNDER (including research centers, academia and industries), the development of JaNDER focused on four key features:

- Modularization of the implementation in order to ensure the integration: RIs with low resources can still implement basic functionality.

- Development of a generic information model as the basis onto which support for more specific information models: this ensures that at least some functionality can be mapped to JaNDER, regardless of the automation level at the individual RI, while contributing to modularization.
- Support for exchange of system configuration information: this supplements the exchange of dynamic data with static data, such as grid topology.
- New replication mechanism: this removes the requirements for opening firewall ports at each participating RI, and the associated administrative overhead.

The ERIGrid JaNDER platform is based on a three-level architecture: this is actually useful both for modularity and flexibility, as well as being open to future extension via additional levels.

2.2   Basic Data Sharing via JaNDER-L0

JaNDER-L0 implements the base functionalities used by all the other layers and is therefore a fundamental building block for the whole architecture. In particular, its main purpose is to allow a basic mechanism for exchanging live data (i.e. typically measurements and controls) between different RIs.

The starting point for each RI is a real time repository based on Redis which is open source, in-memory data structure store, used as a database, cache and message broker. This real time repository is used to collect measurements and controls from the field (or more frequently, as shown in Fig. 3, from an already existing SCADA system): the reason for adopting this repository is decoupling the JaNDER platform from any specific automation solution already installed in the infrastructure. The idea is to have data points from each RI available in the same basic format by using a simple key-value repository. The remote connection of remote infrastructures is then implemented by deploying a common real time repository (which can be hosted in a cloud environment, for example) which is automatically synchronized with all the local real time databases of the partners. In other words, the common repository acts as a central broker for connecting the different local repositories of the partners and can be thought as a "virtual bus" connecting all authorized facilities. There is no handling of standardized protocols or complex interaction patterns above the exchange of data points through the virtual bus.

The fully open source nature of JaNDER-L0 makes it easy to extend the virtual research infrastructure community to new participants. However external users such as other research centres, academia or industries will also typically be interested in having a standardized protocol for interfacing: this is handled by the higher JaNDER levels.

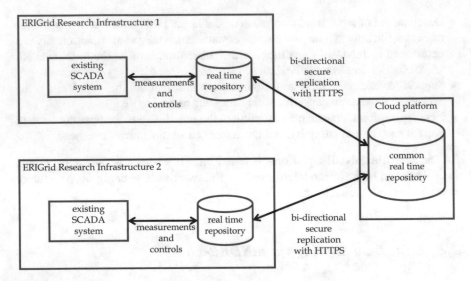

Fig. 3 JaNDER Level 0 architecture

2.3 IEC 61850-Based Communication Platform via JaNDER-L1

JaNDER-L1 is a software abstraction built on top of level 0 and its purpose is to provide an IEC 61850 interface on top of the very simple data structures defined in Redis.

The "Mapping" and "CID" files shown as inputs in Fig. 4 are the fundamental inputs needed by the IEC 61850 server in order to work. More in detail, the CID (Configured IED Description) is the standard IEC 61850 file used for configuring a device (an IED) and contains a data model representing (a subset of) the contents of the Redis repository in terms of IEC 61850 Logical Nodes. Apart from this file, it is of course necessary to link the data attributes defined inside it with the live values stored in Redis: this is done by means of a mapping file, which is a text file where each line contains an IEC 61850 data attribute name and a corresponding Redis data point name. The server will use this file in order to connect the IEC 61850 data model specified in the CID to Redis.

The implemented IEC 61850 connection is always local to the client (i.e., the IEC 61850 client actually runs also the server interface, on behalf of the real information producer) so that cyber security concerns are eliminated at this level.

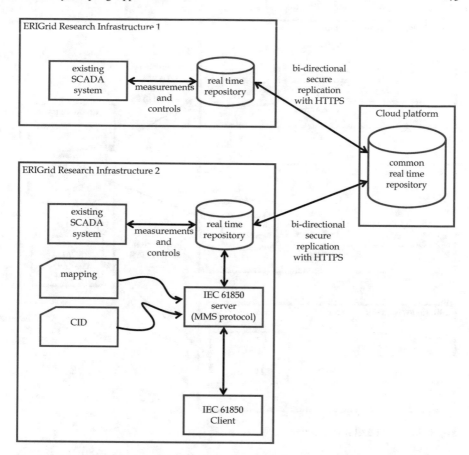

Fig. 4 JaNDER Level 1 architecture

2.4   CIM-based Communication Platform via JaNDER-L2

JaNDER-L2 is a software abstraction build on top of Level 0 (and not Level 1, even if this would be technically possible in principle) and its purpose is to enable the definition of a CIM-based service-oriented architecture on top of the basic live data exchange made possible by the lower JaNDER levels.

The client application (RI3), as indicated in Fig. 5, is an open source graphical interface for handling CIM network representations in conformance with the CGMES profile called CIMDraw, augmented with SCADA interfacing code developed specifically for ERIGrid. Apart from this SCADA interface, which is a different representation of the contents of the real time repository, the main interest for this level lies in the possibility of integrating with other CIM-based services like for example power flow calculation engines, state estimators, voltage control algorithms etc. which can take CIM representations, along with associated measurements, as inputs and produce calculated results as output.

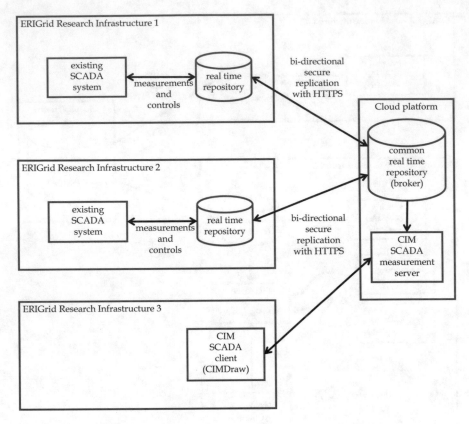

Fig. 5　JaNDER Level 2 architecture

3　Integrated Research Infrastructure

This section provides a description of new power system testing approaches enabled by the communication platform explained in Sect. 2. The demonstration of these approaches has been reported in Sect. 4.

3.1　Hardware/Software Integration Between Different Laboratories

Smart Grids require advanced functionalities in order to optimize their operation. Moving from simulations to actual field implementations could lead to different operational behaviour or could even jeopardize the system's operation. This means that a software or a controller must be tested in relevant environment before the field testing. However, sometimes software developers could require a remote test in order to avoid intellectual property issues. Running the software in their own premises and exchanging data online with another RI where there are monitored, and eventually, controlled devices allows the software developer to protect their know-how. Indeed, not even an executable file is provided to the RI with the hardware. In this case the RI hosting the system under test (at the exception of the software/controller under test) sees only the input and output of the object under test such as a black-box.

3.2 Virtual Research Infrastructure

In order to exploit the synergies among the RIs, each one with its own characteristics, a laboratory coupling is needed. In particular, using the devices of different RIs at the same time enables the extension of the system under test without any further investment in new components. Extending hardware resources of a specific RI by using resources of other RIs allows to implement more test cases than a single RI.

The Virtual Research Infrastructure (VRI) is a combination of RIs coupled by means of a communication platform which combines them in an equivalent bigger laboratory. Hence, in this way a remote hardware can be integrated as a part of the system under test. The interconnection of the RIs avoids additional investments in new hardware and encourages sharing the components of the integrated RIs. The integrated research infrastructure created helps to test components in a real system behaviour also in RIs without a HIL capable simulator.

The technical possibility of conducting such joint experiments allows the application of control algorithms running in one research infrastructure for the remote control of devices which are physically located in other facilities. The advantage of the VRI concept is the possibility for one RI to access the resources located at a remote site - these resources can range from actual hardware devices to real time simulators or Supervisory Control and Data Acquisition (SCADA) systems. A typical VRI can be seen in Fig. 6. In this particular case one of the RIs (TUD) acts as a network simulator while other two RIs (DTU and VTT) participate in a closed loop experiment with their hardware resources.

The main objectives of setting up such an integrated RI is to enable new smart grid testing in a cost-effective solution.

4 Examples of Laboratory Couplings

In this section the demonstration of different approaches cited in Sect. 3, and implemented in ERIGrid, have been discussed. In particular, in order to demonstrate the approaches, the following use case has been taken into account: validation of a centralized voltage control. The implementation of JaNDER enables several test cases for the same use case. The following test cases demonstrate the potential of a commu-

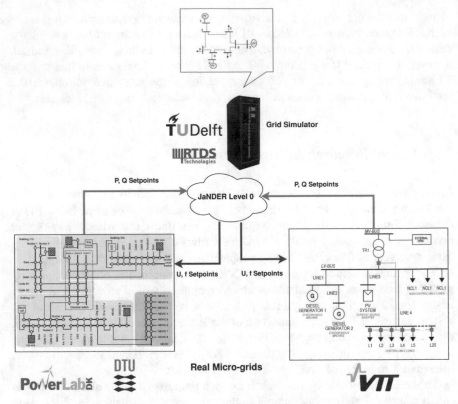

Fig. 6 Integrated research infrastructure involving virtual connection between three research infrastructures

nication platform, which allows to exchange data online between RIs, thus coupling multiple laboratories.

4.1 Integration of a Remote OLTC Controller via IEC 61850

This test case aims at characterizing a software of a RI (ICCS) located in Greece with power system equipment at a remote RI (OCT) located in Spain. The utilization of IEC 61850 through JaNDER-L1 offers the advantage to implement the test case with a widely accepted communication protocol which is used at actual field implementations. In OCT's laboratory the OLTC controller communicates with the local Redis though Modbus protocol. The local Redis updates the cloud Redis which is synchronized also with the local Redis in ICCS RI. Where an IEC 61850 server maps the local Redis measurements and control signals to IEC 61850 logical nodes. Finally, an IEC 61850 client is used to update the control signals in the IEC 61850

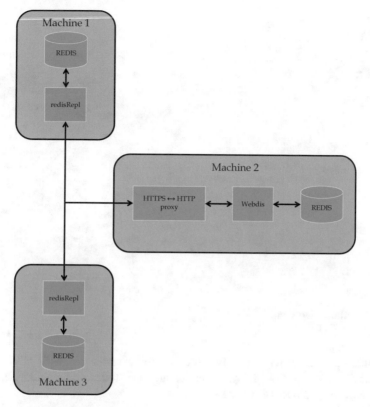

Fig. 7 JaNDER-L1 setup overview

server and acquires the measurements from it. The test case setup is depicted in Fig. 7.

In ICCS RI, a centralized voltage control algorithm (CVC) controls the real OLTC controller located in the UDEX Laboratory within OCT, in Spain. Through JaNDER platform, the controller receives the tap position measurement, performs the optimization and sends commands to increase or decrease the tap position of the OLTC. In ICCS RI, the DRTS is used also to simulate the LV benchmark network. The simulated network consists of 4 simulated PV systems, a Battery Energy Storage System (BESS) and a simulated transformer that changes its tap position according to the tap position signal provided by the OCT's OLTC controller. The controller also receives measurements and sends commands in the simulated LV benchmark network as part of its operation. The test setup is presented in Fig. 8.

In this test case the IEC 61850 interface adds a further delay of 7–8 ms compared to the JaNDER-L0 implementation. This amount of time delay is insignificant for the testing of this kind of controller (CVC), which has a time step of seconds. Therefore, because the tap change is not a delay critical operation, no negative effects have been observed due to these time delays during the experiments. Finally, it is safe

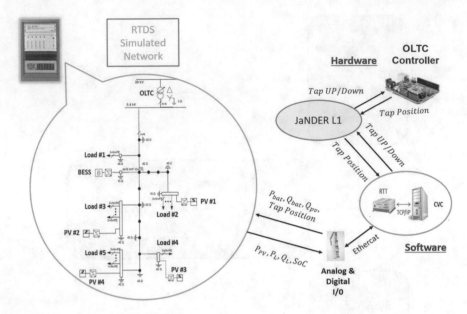

Fig. 8 JaNDER-L1 test setup

to assume that JaNDER-L1 can be used at similar test cases with JaNDER-L0, since the difference in time delays is very small and it relies on the widely accepted communication protocol of IEC 61850.

4.2 State Estimator Web Service

One of the test cases performed in ERIGrid concerns the state estimation via web of a RI that publishes the measurements with JaNDER. In this case the state estimation is done using the Common Information Model (CIM) through JaNDER-L2. In particular one RI on which the web service state estimator has been demonstrated is Tecnalia's smart grid laboratory. On the RI side, JaNDER-L0 was implemented. The system architecture is shown in Fig. 9.

In order to connect the physical devices in the laboratory (Inverter_1, Inverter_2, Load Bank_1, etc.) to JaNDER-L0, a set of communication protocol gateways have been developed. These gateways are software applications in charge of translating the communications between the device specific protocol (Modbus in this case) and the JaNDER-L0 protocol (based on Redis). The gateway applications perform basic tasks such as periodically polling devices for measurement and status values and executing commands published in JaNDER, this is accomplished by mapping Redis variables to Modbus registers and vice versa. The local Redis instance is connected to other applications such as Node Red used for data processing and Redis commander

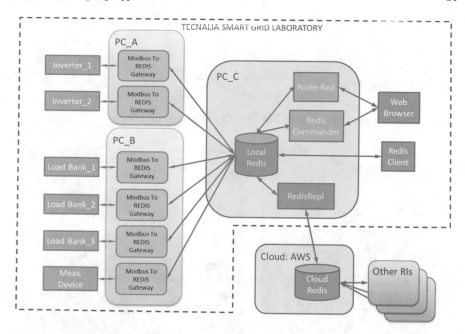

Fig. 9 Implementation of JaNDER-L0 in Tecnalia's smart grid laboratory

for accessing the data through a web client. In addition to this, the local Redis instance is connected to the RedisRepl application in charge of replicating the Redis keys into a Redis remote instance hosted in the cloud. This mechanism allows the integration of devices at Tecnalia's laboratory with other research infrastructures since all the Redis local data can be accessed by JaNDER.

The implementation of JaNDER-L2 is based on the development of a CIM model containing the representation of the laboratory network. This file contains the link to the actual measurements as names of CIM analogue objects which are set to be the corresponding keys in JaNDER-L0. This allows an application using the CIM model to ask for needed measurements to JaNDER-L0 and updating their values on demand. A deployment example is shown in Fig. 10 where the state estimator has been integrated with JaNDER. The user of the state estimator uses the application through a web interface. This state estimator is connected to a Redis instance hosted in the cloud and obtains the real time measurement data that it needs from the Redis cloud instance.

4.3 Geographically Distributed Real-Time Simulation

While the previous two examples show the hardware/software integration between two RIs, the last examples aim at demonstrating the "Virtual Research Infrastructure"

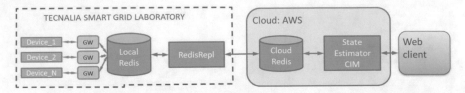

Fig. 10 Implementation of JaNDER-L2 in Tecnalia's smart grid laboratory

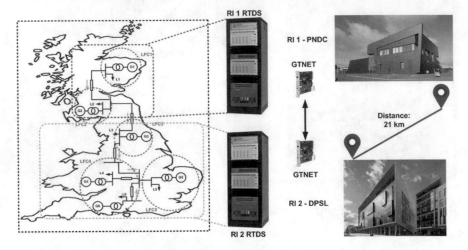

Fig. 11 Geographically distributed real-time simulation setup

Table 1 Summary of GDRTS implementation

Laboratories coupled	Dynamic Power Systems Laboratory (DPSL), University of Strathclyde, UK Power Networks Demonstration Centre (PNDC), UK
Communications	UDP, No orchestrator, Phase compensation, No VPN
Interface and Coupling	Ideal transformer method with synchronous AC coupling
Demonstrated application	Frequency control within a large transmission system

approach. The integration of the RIs can emulate the same setup of a SIL, CHIL or PHIL implementation.

An example of implementation where a large power system is split for simulation within two DRTS at two different laboratories was carried on within ERIGrid. The test setup utilized is shown in Fig. 11 and summarized in Table 1.

The objective of the study was to establish the feasibility of utilising a geographically distributed real-time simulation setup to analyse power systems dynamic phenomena, such as frequency events and incorporating controls at similar time scales. Due to the objective under consideration, JaNDER was not utilized as the communications interface between the two geographically separated laboratories.

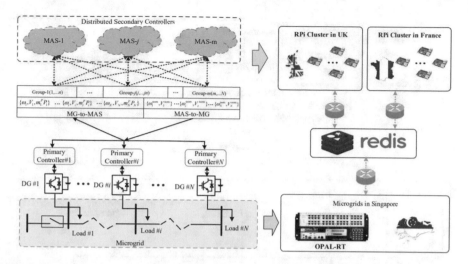

Fig. 12 Real-time geographically distributed controller hardware-in-the-loop setup [4]

Table 2 Summary of RT-GD-CHIL implementation

Laboratories coupled	Dynamic Power Systems Laboratory (DPSL), University of Strathclyde, UK CEA, France Nanyang Technological University, Singapore
Communications	UDP. Web-JaNDER. No compensation, No VPN
Interface and Coupling	Control coupling
Demonstrated application	Frequency and voltage control of microgrids

The results obtained provide confidence in the feasibility of the approach and suggest to carry on further research activities on this topic.

4.4 Real-Time Geographically Distributed CHIL

An example implementation where a power system is simulated within a DRTS at one laboratory while the controller for the power system is implemented within another laboratory was undertaken within ERIGrid. The test setup utilized is shown in Fig. 12 and summarized in Table 2.

The objective of this study in the context of ERIGrid was to establish the feasibility of conducting geographically separated CHIL experiments for power system dynamics studies. The results presented in [4] prove the capability of such setups to undertake real-time power system voltage and frequency secondary control studies.

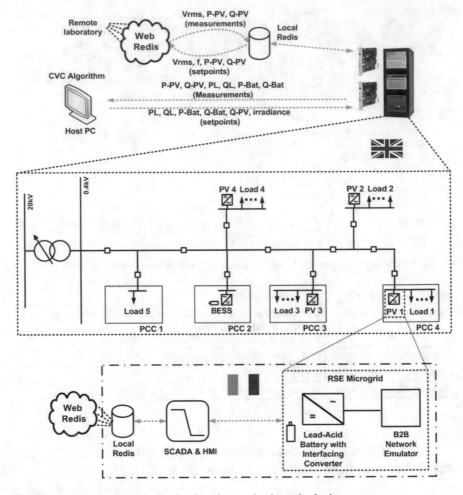

Fig. 13 Real-time geographically distributed power hardware-in-the-loop setup

4.5 Real-Time Geographically Distributed PHIL

An example of inter-laboratory coupling, where hardware resources from two laboratories have been utilized for mutual benefit to enable extended validation capability, is discussed in this sub-section. The utilized test setup is presented in Fig. 13 and summarised in Table 3.

The objective is to validate a CVC algorithm. The test setup involves the simulation of a distribution LV network in real-time within the DRTS at DPSL, where the bus 11 of the network is represented by a lead-acid battery unit at RSE. The voltage magnitude and frequency from the point of common coupling (bus 11 in this case) are sent to RSE via JaNDER-L0 for reproduction within their microgrid

Table 3 Summary of RT-GD-PHIL implementation

Laboratories coupled	Dynamic Power Systems Laboratory (DPSL), University of Strathclyde, UK RSE, Italy
Communications	UDP, Web-JaNDER, No compensation, No VPN
Interface and coupling	Ideal transformer method for asynchronous AC coupling-electrical and control coupling
Demonstrated application	Voltage control of a distribution network

using the back-to-back network emulator. The active and reactive power measured in response to the voltage are measured and sent back to DPSL for injection within DRTS bus 11 using controlled current sources. The CVC algorithm is implemented in a CHIL implementation. The CVC receives the active and reactive powers of the individual buses of the network, processes the inputs to find solution to mitigate any voltage issues identified and determines the new setpoints for the reactive power injection/absorption by the PVs and the active and reactive power setpoints for the BESS.

5   Conclusion

In this chapter different kinds of RI couplings have been presented with the goal of exploiting the synergies among them and making their resources available for external users avoiding additional investment costs. In order to integrate different RIs, a suitable communication platform is necessary. The tool developed in ERIGrid, JaNDER, is able to establish a real-time communication among several RIs. The very low communication latency allows to implement steady-state analysis of the extended system under test. Moreover, JaNDER is able to use also standard such as IES 61850 or CIM. This allows each type of user (e.g. academia using custom protocol, or industries using standard) to access to remote laboratories in a very simple way.

Some examples of test cases performed in ERIGrid using JaNDER concern the software/hardware integration in different RIs, a web service application and a real-time geographically separated PHIL. All these tests were successfully performed and proved that JaNDER satisfies the communication requirements for each test case. The results obtained are very encouraging for the future research activities on smart grid testing. In particular new laboratory coupling could be developed in the future, integrating different kinds of resources and developing a tool to manage the integration in a simple way.

These advanced testing methods could be used to enable new use cases and create a cooperative RI for smart grid system integration which allows the easily transfer of validated solutions into the "real world".

References

1. IEEE standard for interconnection and interoperability of distributed energy resources with associated electric power systems interfaces. IEEE Std 1547-2018 (Revision of IEEE Std 1547-2003) pp. 1–138 (2018)
2. IEEE standard for the testing of microgrid controllers: IEEE Std 2030(8-2018), pp. 1–42 (2018)
3. Nguyen, V.H., Lam Nguyen, T., Tran, Q.T., Besanger, Y., Caire, R.: Integration of scada services in cross-infrastructure holistic tests of cyber-physical energy systems. In: 2019 IEEE International Conference on Environment and Electrical Engineering and 2019 IEEE Industrial and Commercial Power Systems Europe (EEEIC/I CPS Europe), pp. 1–5 (2019)
4. Yu, W., Nguyen, T.L., Nguyen, V.H., Syed, M.H, et al.: A distributed control scheme of microgrids in energy internet and its implementation. Submitted to IEEE Trans. Indust. Inform. (2019)

From Scenarios to Use Cases, Test Cases and Validation Examples

K. Maki, A. Kulmala, K. Heussen, O. Gehrke, E. Rikos, J. Merino, M. Rossi, L. Pellegrino, C. Sandroni, M. Z. Degefa, H. Taxt, D. Lagos, and P. Kotsampopoulos

1 Test Scenario Descriptions

In the context of ERIGrid, scenarios are meant to be higher-level circumstance descriptions which will provide a basis for more detailed use case and test case definitions. As a term, scenario often refers to visionary descriptions of future development and the factors influencing it. Scenarios obviously apply long view perspectives where many uncertainties are present. In the context of ERIGrid, scenarios reaching to 2050 are of interest. In many cases, scenario work can feed in to political processes and decision making on different levels.

In the course of ERIGrid, generic system configurations have been considered more useful than traditional high-level scenarios. A system configuration approach allows including more detailed and quantitative data in the descriptions and providing a better technical basis for developing the use cases and test cases. Whereas high-level

K. Maki (✉) · A. Kulmala
VTT Technical Research Centre of Finland, Tampere, Finland
e-mail: kari.maki@vtt.fi

K. Heussen · O. Gehrke
Technical University of Denmark, Roskilde, Denmark

E. Rikos
Centre for Renewable Energy Sources and Saving, Athens, Greece

J. Merino
TECNALIA Research & Innovation, Derio, Spain

M. Rossi · L. Pellegrino · C. Sandroni
RSE Ricerca Sistema Energetico, Milan, Italy

M. Z. Degefa · H. Taxt
SINTEF Energi AS, Trondheim, Norway

D. Lagos · P. Kotsampopoulos
National Technical University of Athens, Athens, Greece

© The Author(s) 2020
T. I. Strasser et al. (eds.), *European Guide to Power System Testing*,
https://doi.org/10.1007/978-3-030-42274-5_6

scenarios give some qualitative statements about the progress, system configuration uses quantitative data such as numbers of components, size of the system, etc. At the same time, the system configuration becomes more complex due to the amount of data but also more locally due to dimensions and local parameters.

The system configurations allow development of use cases, which give a description of a process leading to a specific objective. In other words, use case defines the actions needed to obtain some goal. Use cases are often described from an external perspective in a neutral manner, utilizing a formal methodology. Use cases can also be thought to define the interfaces of the process with its environment, inputs and eventual outputs.

Use cases can be defined from two perspectives: behavioural perspective and interaction perspective. Behavioural perspective is always function-type; it defines the behaviour of the process internally and towards external stakeholders. In the interaction perspective, most interest is on interactions between components and describing them, for instance by means of sequences.

Test cases with reference to system configurations require information on system parameters, ranges of parameters, system functionalities and quantitative measures. They also require information on test procedures and design of experiments. Test cases define the actual test setup; which are the combinations and series to be tested and which are the prevailing circumstances in which the tests are performed.

Following definitions have been used within ERIGrid [3]:

- *System* defined as a set of interrelated elements considered in a defined context as a whole and separated from their environment.
- *System Configuration* defined as an assembly of (sub-)systems, components, connections, domains, and attributes relevant to a particular test case.
- *Scenario* defined as a compilation of System Configuration, Use Cases, and Test Cases in a shared context.
- *Use Case* defined as a specification of a set of actions performed by a system, which yields an observable result that is, typically, of value for one or more actors or other stakeholders of the system.
- *Test Case* defined as a set of conditions under which a test can determine whether or how well a system, component or one of its aspects is working given its expected function.

2   ERIGrid Generic System Configurations

ERIGrid has defined three system configurations addressing key system areas [3]:

- Distribution grid
- Transmission grid and offshore wind
- Vertical integration

The project has developed system configurations for these dedicated system areas as well as structures and templates for describing them. The templates apply similar hierarchy for the structure, starting from domain information and proceeding to more local information (such as area/level) and finally to individual components. Parameters are defined for each component as well as for the whole system as global parameters.

Distribution Grid

System configuration "Distribution grid" considers the electricity distribution system at MV and LV voltage levels. The area covered by this configuration starts at the HV/MV transformer, where also the responsibility area of DSO typically starts. On the low voltage side, the configuration is limited to the customer interface (metering point) or at the connection point of each active component or DER unit. However, the configuration also needs to consider components beyond the network connection point to the degree they impact on the state of the distribution grid. Hence components like control systems for DER units or controllable loads are included in the configuration.

The distribution grid as a domain includes a significant number of control-related challenges and developments. Communication is also increasingly present for monitoring and control purposes. One issue faced in this work was how to present these different layers. It could be possible to build up separate layers for the power system, communication system and control systems. This would enable a more detailed presentation of each system and especially of their interfaces. Eventually, control systems and ICT have been included as separate domains in this configuration. Multi-domain components are located in domain interfaces, for instance smart meters which are physically connected to the power domain but also connected to the ICT domain in terms of data and control.

This system configuration includes a long list of traditional power system components such as lines, loads, transformers and switches. They all belong to the electrical power system domain. Some active components such as DER units, storage units, EV charging stations or intelligent controllers are also present; they are also physically connected to the electrical power system do-main, but they are also connected to control and ICT domains via their controllers and com-munications.

The system configuration also includes a heat system domain. The purpose of including a heat system is to be able to represent aspects of cross-impacts between heat and electricity; for instance, in a Combined Heat and Power (CHP) production, between heat exchangers, heat pumps, etc. However, heat system parametrisation is left very generic with the main focus to include connectivity.

The control domain includes various controllers connected with components. They have been categorized to central (coordinated) and local control methodologies. ICT domain includes metering systems, communication and data management areas. Stakeholders and markets have also been presented as separate domains, indicating different roles and markets within the scope of this system configuration.

Transmission Grid and Offshore Wind

The offshore wind power plant scenario has been selected because it is a predominant future scenario with special operation characteristics and impact on transmission grids. For specifying the system configuration, the following assumptions have been made:

- A meshed HVDC network will be adopted because it seems a cost-effective solution for hosting high-power wind generation and, as a topic, it presents an additional research interest.
- AC grid parts are assumed for the connections of the wind power plants to the HVDC hubs and an aggregated representation for the on-shore substations/connections.
- More than one connection to the shore may be used because it adds extra benefits in terms of services and allows the wind power plant to participate in various processes of operation and the energy market. Also, this increases the number of applicable use cases.
- Interconnection with different control areas (different countries) so as to increase diversity of operating characteristics and processes at the ends of the system.
- Simple configuration with the minimum possible number of components that at the same time satisfy the abovementioned requirements.
- Hierarchical control structure based on levels, with each level assigned with specific roles for the system's protection, operation and optimisation.
- The system is assumed to have specific role(s) in the energy and ancillary services market which help to establish concrete interconnections with the 'Market' domain.
- The interconnection with other physical domains such as weather conditions is more specific since there is only one RES technology involved. Nevertheless, the effects of weather conditions are considered only as a boundary of the system and are not analytically modelled.

Based on example scenarios, the system configuration is extended according to the aforementioned assumptions. To this end, components given in the basic scenario have been identified followed by components for possible extensions to the basic scenario. For those components, attributes and domains have been identified as well as the connections between.

One of the most crucial discussion topics was the importance of considering onshore wind power plants together with the offshore scenario. The former is (and will be) the predominant wind-production scenario of the future. However, taking into account only the share of a scenario for selecting it, it means that other large-scale technologies should also be considered. Thus, only the offshore wind power plant scenario is considered, not just for its contribution to the RES share but also for its technical characteristics. Specifically, the incorporation of meshed HVDC grids is a value added for the selection of the scenario.

The topology of the system was also an important discussion topic. Among different options such as pure AC, radial DC, and meshed DC configurations, the meshed scenario has been selected which is technologically the most promising solution for bulk transmission of offshore wind power.

HVDC onshore fault ride-through protection was also identified as a serious challenge from an operational standpoint, as well as from testing and simulation perspectives.

A third point of discussion was the way of modelling the onshore connection points and, in general, the overall onshore transmission grid's behaviour in combination with the selected scenario. To this end, aggregation of production/consumption at various grid nodes (at transmission level) and simplified representation of the transmission grid has been agreed. With the use cases in mind (e.g., fault ride-through, energy balancing, active power control, stability to a lesser extent) this is a plausible assumption.

Vertical Integration
The vertical integration scenario and system configuration provides a possible background for use cases requiring coordination and integration of transmission and distribution grid related tasks. In principle, it includes all domains used in other system configurations; however, in this system configuration often abstractions and aggregations of usually included components are employed, as the full detail may overload a given test requirement.

Due to its cross-cutting nature, vertical integration system configuration sets a lot of attention on connectivity of components, their information exchange as well as on the roles of stakeholders.

3 Focal Use Cases

The ERIGrid Focal Use Case Collection has been gathered during the project, based on existing outputs from earlier projects and networks. Several repositories, for instance EPRI (The Electric Power Research Institute) and SGCG (Smart Grid Coordination Group) ones have been utilized while building the ERIGrid collection.

Focal use cases have been categorised according to the service they provide for the system [4]:

- SS1 Energy balance
- SS2 Energy efficiency
- SS3 Power quality
- SS4 Power system stability
- SS5 Infrastructure integrity, protection and restoration

The following sub-sections present these services that can be provided at system level and show exemplary use cases for each of them. These use cases are aggregates of several use cases within the ERIGrid collection.

SS1 Energy Balance The energy balance of a network is a fundamental requirement for its operation; in fact, the generation has to constantly follow the demand curve in order to maintain the system stable. Taking into account the time horizon for which the ERIGrid system configurations have been developed, scenarios are included in which also the demand is controlled in order to match the generation availability. Here the energy balancing functions are defined as follows:

- Functions aimed at guaranteeing the long-term energy balancing and which have been categorized in SS1 Energy balance in the restoration of the planned power exchanges with external systems.
- Functions aimed at guaranteeing fast and prompt support in the restoration of the power balancing have been categorized as Focal UC in SS4 Power system stability.

The focal Use Cases listed below describe the selected functions for the support of system energy balance:

- SS1.SC1 Management of Flexible DERs for the Long-term Balancing (Frequency/Voltage Restoration Reserve) of Microgrids in Island-Mode
- SS1.SC2 Automatic Frequency Restoration Reserve from VSCs of Large Wind Farms
- SS1.SC3 Automatic Frequency Restoration Reserve from DERs

SS2 Energy Efficiency The containment of the losses in energy conversion, transportation and storage has always been one of the main objectives in the design and operation of power systems. In fact, several use cases can be found or deduced from literature specifically aimed at enhancing the energy efficiency of systems. Energy efficiency related use cases have high relevance among the ERIGrid collection:

- SS2.SC1 Optimal Distribution Network Control for the Reduction of System Energy Losses
- SS2.SC2 Optimal Transmission Network Management Level for System Energy Losses Reduction
- SS2.SC3 Incentivising Distribution Network Local Balancing to
- Minimize Transmission Network Loading

SS3 Power Quality The increasing penetration of distributed generation is particularly challenging from the power quality point of view and, currently, one of the most relevant limitations in terms of renewable integration are the voltage issues caused by generation at distribution level. In order to mitigate these effects, potential solutions have to be developed and most of them require the coordination of more resources in order to manage the voltage congestions. According to this, all the ERIGrid system configurations can be considered as proper scenarios in which power quality functions can be tested and, for each of them, specific focal use cases are listed:

- SS3.SC1 Advanced Voltage Control of Distribution Grids Supported by DERs Power Interfaces
- SS3.SC2 Voltage Quality Support by Onshore and Offshore (VSC-HVDC connected) Wind Power Plants
- SS3.SC3 Transmission Network Voltage Quality Support by the Distribution Network (VPP)

SS4 Power System Stability Another particularly challenging aspect in future power systems is represented by stability. Most of the functions aimed at supporting the system robustness are currently performed by traditional generators. For scenarios in which their presence is expected to be less predominant, other solutions have to be exploited. In order to support the power system stability, many actors can be involved as well as all the domains considered within ERIGrid. Also in this case, on the basis of ERIGrid System Configurations, a list of use cases is presented:

- SS4.SC1 Management of Flexible DERs for the Instantaneous Active/Reactive Power Balancing of Microgrids in Island-Mode
- SS4.SC2 Large-scale Wind Power Plant (Onshore and Offshore VSC-HVDC Connected) Support in Frequency Containment Control and Power System Inertia
- SS4.SC3 DERs Support in Frequency Containment Control and Power System Inertia

SS5 Infrastructure Integrity, Protection, and Restoration Other functions that are expected to evolve in the ERIGrid system configurations are represented by the ones supporting the integrity, protection and restoration of the System Configurations' infrastructures. In fact, taking into account the high flexibility that energy players are able to provide at all power system levels, significant benefits can be provided through theses dedicated use cases. As for other services, use cases are listed, including also a use case describing functions aimed at guaranteeing ICT integrity, protection and restoration:

- SS5.SC1 Fault Detection and Corrective Management of Distribution Grid Assets and Energy Resources
- SS5.SC2 VSCs (of HVDC and Large Windfarms) Support During Transmission Network Restoration
- SS5.SC3 Intentional Islanding of Microgrids During Widespread Disturbances and Restoration of the Transmission System
- SS5.SC4 Identification of ICT Anomalies and Restoration of the Communication Links

These listed sixteen focal use cases have been considered representative of most relevant functions that can be reasonably expected to be operative in the ERIGrid system configurations. These focal use cases have been designed in order to cover a large spectrum of system domains and actors, and to comprehend several more specific functions. Based on this, several test cases can have been designed, taking advantage of the different domains which can be easily reproduced and/or simulated within ERIGrid research infrastructures.

4 Test Cases

ERIGrid covers all testing approaches consisting of virtual-based and/or real-world-based methods. The test set-ups can be divided into four categories: (i) pure simulation (incl. co-simulation), (ii) Controller Hardware-in-the-Loop (CHIL) simulations, (iii) Hardware (HW) experiments, and (iv) Power Hardware-in-the-Loop (PHIL) experiments. These testing approaches can be considered to form a structured testing chain consisting of the following steps:

- Pure simulation: Virtual-based approaches both offline and in real-time. All aspects of the System under Test (SuT) are modelled using suitable software(s) and the accuracy of the results depends on the accuracy of the utilized models. Co-simulation can be used to combine different simulators that consider each domain-specific part of the SuT individually.
- CHIL experiments: Real control hardware is utilized in a closed-loop simulation of the system. Virtual-based and real-world-based approaches are combined. CHIL experiments enable more accurate simulations in case an exact model of the controller is not available. Communication delays, noise, execution time of algorithms etc. can be taken into account more easily than with a pure simulation approach. CHIL experiments can also be used to verify the correct operation of a specific control hardware.
- Hardware experiments: Open-loop testing of real components. This can be seen as the conventional part of component testing and the results are mainly related to component characteristics.
- PHIL experiments: Closed-loop testing of real components. Virtual-based and real-world-based approaches are combined. Interactions between the hardware under test and the overall system can be studied.

The next step after the four testing approaches would be demonstration in a real operational environment. The test cases are selected to cover all of the four testing approaches. Better RI integration is needed to enable comprehensive testing of multi-domain systems and also to enhance the already existing single-domain testing procedures. Better integration can be achieved by at least two means: By simplifying the process of porting an experiment from one RI to another, e.g., by using standardized interfaces, and by enabling joint use of RIs with different capabilities through a real-time communication between the RIs. Here the concept of an RI is understood to include off-line simulation tools as well as physical laboratory infrastructure consisting of real equipment such as generators and virtual equipment such as real-time simulators. Research questions determine the test objectives for each of the selected test cases. Additionally, previously defined system configurations and focal use cases as well as capabilities of different RIs are used as inputs for the selection process. The number of tested use cases is intentionally quite low so that the work can concentrate on the research questions on infrastructure integration. Same use cases are used for many different test cases so that the testing approaches can be developed and results of individual setups can be compared.

Selected Test Cases

The work concentrated on demonstrating and validating research infrastructure integration. Two types of test cases have been defined: In single-RI integration test cases, models, algorithms etc. developed in one RI are used in another RI as a part of a test but real-time communication between the RIs is not needed. Single-RI test cases are also used to compare different experiment set-ups to enable performing the same tests in different facilities. In multi-RIs integration test cases, the interfacing and real-time communication between the RIs is needed.

Technical challenges identified for single-RI test cases include comparison of different testing approaches and test setups, integrating third-party Software (SW) as a part of a test case and model transfer between RIs. Four single-RI test cases have been selected as presented below. The following test cases have been selected as single-RI test cases [2]:

- TC.S.1 Component testing at different RIs with different setup
- TC.S.2 Use of SW developed by RI1 in HW RI2
- TC.S.3 Use of component model developed in RI1 to perform multi-domain system tests in RI2
- TC.S.4 Test of distributed cyber-physical systems in RI1 as a monolithic setup in RI2

Technical challenges identified for multi-RI test cases include integration of remote software and integration of remote simulators or hardware. The multi-RI test cases are used both to validate the correct operation of interfaces developed and to demonstrate the real-time joint operation of research infrastructures for smart grid testing purposes. Two multi-RI test cases have been selected as presented below. The following test cases have been selected as multi-RI test cases [2]:

- TC.M.1 Integrate control SW running in RI1 with HW RI2
- TC.M.2 Extend HW resources of RI1 using resources of other RIs

More detailed implementation plans for each test case have been developed, started by constructing test specifications and experiment specifications as well as experiment setups. Full test case descriptions include system configurations as a basis and requirements for research infrastructures. System configurations provide information on the required components and connectivity.

5 System Validation Examples

The following sections present two more detailed examples utilizing the procedure. The first one deals with a voltage control application and the second one with the development of a converter controller.

5.1 Analysis of the Centralized Voltage Control for Rhodes Island

This test case aims to demonstrate the "Hardware/Software integration between different Research Infrastructures". This test case aimed to demonstrate how modern advanced testing techniques, such as CHIL, can be used to fill the gap and ensure faster and more secure transition between pure simulations and field implementations. In this test case, a control hardware in the loop setup was used in ICCS RI in order to test a DSO's control algorithm of a Non-Interconnected Islanded Power System (NIIPS) in realistic conditions. A dynamic model of Rhodes Island system implemented in ICCS's real time digital simulator in full detail consisted of:

- Synchronous Generators and their control systems (automatic voltage regulators, governors, secondary frequency controls) of the 2 different power stations in the island.
- The HV network of the Rhodes Island system.
- The 5 controllable WTs that exist in Rhodes network (as average P, Q models).
- 5 average P, Q models that represent the demand of the 5 different HV/MV substations of the island.

The dynamic model of the island power system was simulated in real time, sending also measurements and receiving setpoints from a controller hosting DSO's algorithm which operates in a CHIL setup as presented in the next figure. The first control algorithm measures the production of the synchronous generators (thermal units), the power produced by the WTs and the available power of the WTs. From those measurements the total demand is derived. The controller then determines the maximum production allowed cumulative by all the WTs according to their maximum available power, the maximum permitted penetration level that is set for stability purposes (e.g. 30% of the total demand) and the non-violation of the minimum loading levels of generators. The algorithm then decides how to distribute this available power according to the nominal rating of each one of the 5 WTs as well as their respective available power at that moment. Finally, the setpoints for the thermal generators calculated by trying to reduce the production cost while at the same time supply the remaining demand and provide the required reserves for the safe operation of the system. Those setpoints sent alongside with the WTs setpoints back to the RTDS. Furthermore, this CHIL setup is ideal to examine the behaviour of the existing algorithm as well as a possible improvement of it in real time conditions (e.g. noise, time delays) and also in events that can cause stability issues in the system such us the largest WT disconnection. In order to illustrate this, an improvement of the DSO's centralized control tested also in ICCS CHIL testbed and compared to the existing controller. The proposed control algorithm also tries to reduce the voltage deviation from the nominal value, similar to the CVC, as well as to ensure stability due to RES penetration levels through a different approach. The controller hosting this algorithm receives measurements (WT Power, Thermal Unit Production, WT available power, active and reactive power demand of the 5 HV/MV substation) from RTDS, solves an optimization problem and sent back in RTDS the WT active and reactive power setpoints and the thermal generator setpoints. The optimization problem tries to minimize the voltage deviation and the operating costs according to the following objective function:

$$min_x \left\{ w_{cost} \cdot \sum_{i=1}^{\infty} Cost_i \cdot P_g + w_v \cdot \sum_{j=1}^{n} (V_n - V_j)^2 \right\}$$

Subject to the constraints:

- Power Balance equations in each node
- Voltage constraints ($V_{min} \le V_j \le V_{max}$)
- Angle Constraints ($-180° \le d_j \le 180°$)
- Thermal Generator Production Limits ($P_{min} \le P_i \le P_{max}$)
- WT power constraint according to the available power ($P_i^{WT} \le P_i^{WT_{avail}}$)
- WT power factor constraint ($Q_i^{WT} \le P_i^{WT} \cdot \tan \arccos 0.9$)
- Dynamic Frequency Constraints ($F(H, P_{dis}) \le 49.4$)

The last constraints are a set of linear constraints ensuring that the frequency will not drop below a frequency level (here 49.4 Hz which is the setting of the first acting Under Frequency Load Shedding (UFLS) relays and is a recorded transient that has caused black out in the Rhodes NIIPS). This is a different approach compared to the existing control which enforces a penetration limit to the WT power. Both methods try to ensure that if a contingency occurs (e.g. the largest WT is disconnected) the frequency remains within limits and no UFLS relays trip. The CHIL setup in ICCS infrastructures allows to run in the loop both algorithms and perform at the same time contingencies in RTDS in order to examine if the existing and the proposed control algorithm ensure safe operation. In addition, the CHIL allows both control algorithms to be tested in realistic conditions, ensuring that time delays and noise on the signals does not affect the stability of the system (e.g. introduce oscillations in power which also introduce oscillations in frequency). The DSO's proposed control achieved better results compared to the existing control in terms of voltage since it utilizes the ability of the WT's power electronics to absorb or produce reactive power in order to mitigate the voltage issues. The voltage profiles of 2 HV/MV substations in the HV side are presented in Figs. 1 and 2.

The resulting reactive power profiles of the 5 WTs of the second method are presented in Fig. 3.

Furthermore, the proposed control achieved higher penetration levels compared to the existing control as presented in Fig. 4.

This was mainly implemented by dispatching differently the WTs, utilizing more the WTs that have a lower rating if it is not secure to further increase the production of the WTs that a have higher rating. The profiles of the production of 2 WTs are presented in Figs. 5 and 6.

It is observed that the proposed control achieves lower production only in WT2 which has the highest nominal rating and results in higher production levels according to the existing DSO control. The proposed control reduces the power production of this WT in order to avoid a frequency transient that would result in a frequency nadir below 49.4 Hz. The existing CHIL setup allows also to perform such transient in the RTDS in order to compare the two algorithms according to the frequency transients that could occur if any of them is under operation. In Fig. 7 the recorded transients in frequency for the disconnection of the largest producing WT were performed and recorded in different hours of the day proving the superiority of the proposed method.

To sum up, this test case allowed a second party (DSO) to test its control algorithms in an advanced testing setup of CHIL, not available at the DSO's premise, provided by a second party (ICCS). In this setup, the comparison was made in realistic conditions (noise, time delays) and for complex scenarios (WT outages) that could assist the DSO to evaluate and compare those 2 methods.

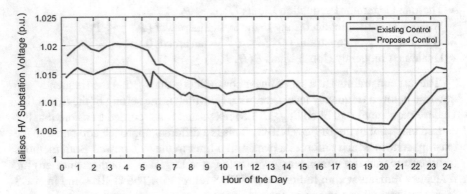

Fig. 1 Comparison of Ialisos HV profile

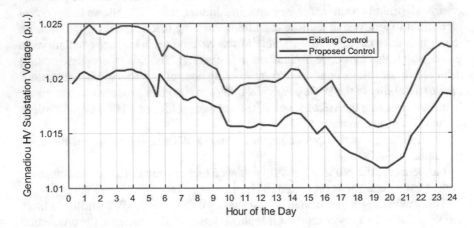

Fig. 2 Comparison of Gennadiou HV profile

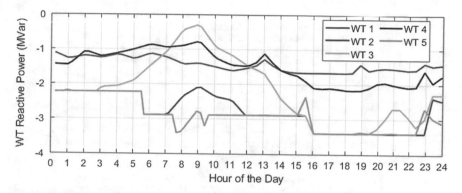

Fig. 3 Reactive Power Profiles throughout the day for the proposed HEDNO control

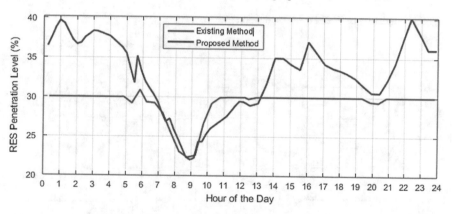

Fig. 4 Comparison of RES penetration levels for both controls

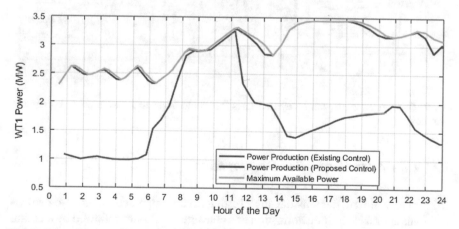

Fig. 5 Comparison of WT1 Power Production profiles

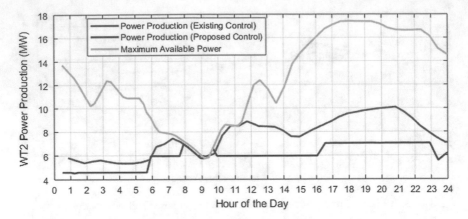

Fig. 6 Comparison of WT2 Power Production profiles

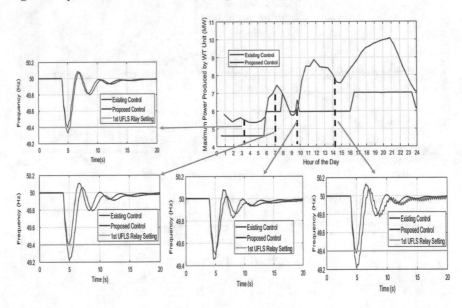

Fig. 7 Comparison of the frequency transients that could occur under each control algorithm operation

5.2 Converter Controller Development

This test case aims to demonstrate the "Testing Chain" approach. In order to demonstrate this approach, a characterization of a converter controller followed by a tuning was done. The test system includes: distribution LV grid, converter controller, PV, inverter. To characterize the converter controller a pure simulation test was implemented firstly, then after a tuning of the converter controller based on the simula-

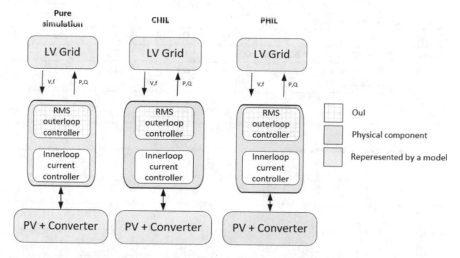

Fig. 8 System under test of the testing chain test

tion results, a CHIL and a PHIL test were performed to validate the improvements. Figure 8 shows the system under test taken into account in each step of the testing chain.

The selected LV test grid, considered as relevant enough for testing the converter controller, is based on the CIGRE LV network modified with DER that can be found in [1]. The object under test is a droop controller that is used to control the PV source in the grid. In order to compare the test results four KPIs have been defined and evaluated:

- Settling time (ST): Time elapsed from the application of an instantaneous step input to the time at which the amplifier output has entered and remained within an error band of 5%.
- Overshoot (OS): $OS(\%) = (V_{peak} - V_{SS})/V_{SS} \cdot 100$
- Time of peak (Tp): Time at which the peak value occurs.
- Damping factor (DF):

$$\theta = \frac{ln\left(\frac{OS}{100}\right)}{\sqrt{\pi + ln^2\left(\frac{OS}{100}\right)}} \quad \begin{cases} \theta < 1 & \text{Under damped} \\ \theta = 1 & \text{Critically damped} \\ \theta > 1 & \text{Over damped} \end{cases}$$

Following a description of the main results achieved with the testing chain.

- **Pure simulation test**:

To cover a wide spectrum of possible operating conditions, several experiment specifications have been defined, for analysing the converter response in case of generation and load variations. The experiment results are in Table 1.

Table 1 Pure simulation results

Test type	Action	ST [s]	OS [%]	Tp [s]	DF
Simulation	Step up PV output of 1 p.u.	0.35	4.34	0.22	0.7
Simulation	Step up of the grid load or generation of 1 p.u.	0.16	0.9	0.15	0.83

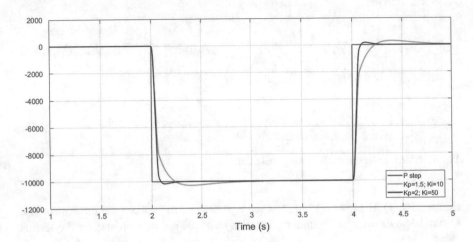

Fig. 9 Response of two [Kp, Ki] sets of parameters

Based on these results the parameters of the converter controller were tuned further. Regarding the inner loop, a parametric analysis has been carried out to evaluate the impact of the PID constants in the system response when facing a step in the solar power generated. In any case, a wide range of values has been tested to find the better trade-off between the settling time and the overshoot. Two sets of [Kp, Ki] constants were selected. Figure 9 shows the new response of the converter controller compared to the old values of set [Kp, Ki].

In view of the results a final tuning of Kp = 2 and Ki = 50 has been eventually selected for the second round of tests.

Regarding the outer loop, potential improvements are linked to the adjustment of the measurement filters in the Id and Iq, and being more concrete, in the adjustment of the damping ratios (D). The damping ratio is a parameter linked to the quality of the filter in the way a higher damping ratio means a higher quality. Also, in this case several values of D have been evaluated in order to establish the best solution. The experiment's results show that the most suitable value for the D parameter of the Iq filter is 0.5 while the recommendation of the D value for the Id filter for the second round of tests is 0.85.

Table 2 CHIL results

Test type	Action	ST [s]	OS [%]	Tp [s]	DF
CHIL	Step up PV output of 1 p.u.	0.02	0.73	0.1	0.94

Table 3 PHIL results

Test type	Action	ST [s]	OS [%]	Tp [s]	DF
PHIL	Step up PV output of 1 p.u. with the original converter controller	0.0084	16	0.0012	0.49
PHIL	Step up PV output of 1 p.u. with the improved converter controller	0.0062	22	0.0014	0.43

- **CHIL test**:

Both sets of controller parameters which were discussed in the previous paragraph have been tested also with a CHIL setup. The results with the original converter controller are in Table 2.

With the CHIL experiments, the improved controller shows that the ST of the system is lower than the original controller, but it has a very oscillatory behaviour. For large steps in active power injection, such as the one required by this scenario, the controller became unstable. Therefore, it is expected that the second version of the controller will behave worse also in reality (in terms of oscillatory response) than the first version of the controller.

- **PHIL test**:

The last step of the testing chain is the PHIL experiment. In this case the converter controller was implemented on a real power converter. Similar to the CHIL experiment, also in this case the results show that the improved converter controller reduces the TS at the expense of the level of the OS (Table 3).

6 Conclusions

This chapter described the progress from general scenario thinking towards more detailed system configurations, use cases and test cases. Validation examples were presented to demonstrate the usage. The main objective was to access the relevancy of testing needs for smart grid system development. The work has started with the development of generic system configurations which provide a high-level context for ERIGrid testing. The work has proceeded to define focal use cases which cover the whole range of smart grid activities relevant to ERIGrid. Based on focal use case collection, most relevant testing scenarios for have been defined on an abstract level, outlining the most important areas of smart grid testing. The application examples demonstrate usage of the process within different circumstances. A first example demonstrates hardware/software integration across research infrastructures. In this case, CHIL and dynamic modelling were used as testing techniques. The second example focuses on testing chain approach, progressing from simulation studies to CHIL and PHIL tests.

References

1. Kotsampopoulos, P., Lagos, D., Hatziargyriou, N., Faruque, M.O., et al.: A benchmark system for hardware-in-the-loop testing of distributed energy resources. IEEE Power Energy Technol Syst J **5**(3), 94–103 (2018)
2. Kulmala, A., Maki, K., Sandroni, C., Pala, D., et al.: D-JRA1.3 Use case implementation plan. Deliverable D7.3, ERIGrid Consortium (2017)
3. Maki, K., Blank, M., Heussen, K., Bondy, E., et al.: D-JRA1.1 ERIGrid scenario descriptions. Deliverable D7.2, ERIGrid Consortium (2016)
4. Rossi, M., Carlini, C., Pala, D., Sandroni, C., et al.: D-JRA1.2 Focal use case collection. Deliverable D7.2, ERIGrid Consortium (2016)

Experiences with System-Level Validation Approach

P. Teimourzadeh Baboli, D. Babazadeh, D. Siagkas, S. Manikas,
K. Anastasakis, and J. Merino

1 Introduction to Users and Experiences

ERIGrid integrates and enhances the necessary research services for analysing, validating and testing smart grid configurations. System level support and education for industrial and academic researchers is provided as well to foster future innovation. To this end, some internal and external programs have been defined to use the available research infrastructures provided by ERIGrid. From 2016 to 2020, ERIGrid enabled more than 100 research projects, in which most of the research teams used the Holistic Test Description (HTD) approach [1] for documentation of the test procedures and results as an effective *system-level validation* approach. This pool of HTD experiences created a valuable potential to collect feedback, validate the methodology and therefore presented a chance to refine the methodology. In this chapter, the application areas of the HTD methodology are either used internally in ERIGrid or adopted for external usages as discussed. The evaluation of the system-level validation is addressed in and a summary of the HTD users' experiences are explained. Also advantages and shortcomings of the HTD methodology based on the integrated feedback of the users are presented.

P. Teimourzadeh Baboli (✉) · D. Babazadeh
OFFIS – Institute for Information Technology, Oldenburg, Germany
e-mail: payam.teimourzadehbaboli@offis.de

D. Siagkas · S. Manikas · K. Anastasakis
Hellenic Electricity Distribution Network Operator, Athens, Greece

J. Merino
TECNALIA Research & Innovation, Derio, Spain

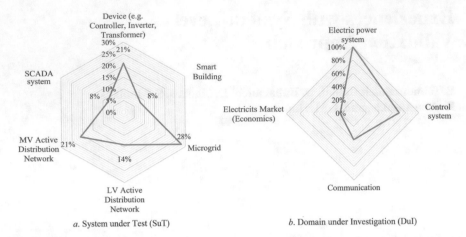

a. System under Test (SuT)   *b.* Domain under Investigation (DuI)

Fig. 1 Analysed **a** Systems under Test (SuT) and **b** Domains under Investigation (DuI)

2 Application of System-Level Validation Approach in Projects

As mentioned in Chap. 2, the HTD method has been proposed in the ERIGrid project as a template-based documentation and system-level validation approach. To have a better picture of the HTD application experiences, they are clustered based on the *System under Test* (SuT) and *Domains under Investigation* (DuI) aspects. As shown in Fig. 1a the SuT of the projects has been categorized in six main clusters; namely: *(i)* Device-level testing, *(ii)* smart building, *(iii)* microgrid, *(iv)* low-voltage distribution network, *(v)* medium-voltage distribution network, and *(vi)* Supervisory Control and Data Acquisition (SCADA) system. Most of the HTD users chose the *microgrid* as the SuT. *Device-level testing*, such as evaluating the performance of voltage controller, inverter and transformer, is the second most popular category.

Figure 1b represents the different DuI of the HTD users, which are categorized in four main domains; namely: *(i)* electric power systems, *(ii)*control systems, *(iii)* communications, and *(iv)*electricity markets. The total percentage of DuI percentage exceeds from 100% due to the domains overlap in the projects. So far, all the users studied power systems as the main domain with different focuses on control, communication and market perspectives.

3 Evaluation of Representative Test Cases

The application of the HTD which is the proposed system-level validation approach in the ERIGrid project, is analysed for two different test cases involving *a single* or *several joint infrastructures*. The results of this analysis are presented in this section.

Single research infrastructure test case: A converter controller testing has been validated following a complete testing chain: *(i)* pure simulation, *(ii)* Controller Hardware-in-the-Loop (CHIL), and *(iii)* Power Hardware-in-the-Loop (PHIL). The objective of this test case is to accomplish several experiments aiming at evaluating the impact of the control and the parameters of the controller in the performance. For that, the different droop controls (P-f and Q-V) have been characterized and the results are compared. The test cases have been formulated based on three main Purpose of Investigations (PoIs):

- Characterization of the converter controller influence of the system performance (i.e., fine tuning of models using results of hardware tests).
- Validation of model exchange among RIs.
- Validate improved control system performance.

This test case with several PoIs becomes a complex example of the use of the methodology. Therefore, the application of the HTD was done for one purpose of investigation considering several sub-PoIs. Due to the specific requirements of the tests to be performed, the design of an experiment realisation plan becomes challenging as the information provided by the labs is usually not so detailed to ensure replicability across different research infrastructures.

Several joint infrastructures test case: The second evaluated example was the test of a web service provided by one research infrastructure using data from another research infrastructure. Its aim is to validate a software available as a web service by using the measurements registered in one research infrastructure and sent to the cloud using a certain protocol, i.e., Common Information Model (CIM) [2, 3]. This test case has two PoIs related to the validation of:

- A state estimator web service.
- A virtual research infrastructure.

From the formal point of view, the application of this example is somehow incomplete. Moreover, evidences in different fields of the template may lead to misunderstanding on the whole procedure. One of the most outcomes and lesson learned from the application of this example is underlining the importance of defining common definitions and a clear guideline for the application of the methodology.

4 Evaluation of the Holistic Test Description Methodology

A questionnaire has been designed and handed out to the people employing the ERIGrid services associated with the holistic testing and validation procedure in the context of test cases in different projects. The goal of this questionnaire is to document issues and shortcomings of the services, in order to improve them and document the iterative development process. The following services are addressed by this questionnaire:

- *Holistic test case description:* Templates and possibly accompanying guideline material for documentation of test case, test specifications and experiment specifications.
- *System configuration description:* Filling forms for documentation of the test system configuration as part of the holistic test case description.
- *RI Database:* Web-based platform for formal RI component specification.

The Data Specification Questionnaire is directed at all researchers who have employed parts of the HTD procedure. The goal is to point out the improvement potential of the procedure, especially in the context of data specifications.

4.1 Results of Work with ERIGrid Services Questionnaire

From the collected answers the majority of the TA users who used the holistic test case description service, used the system configuration description service, and only a few of them used the RI database service, which is the web-based platform for formal RI component specification.

The templates seem to be quite clear and convenient. Most of users did not encounter any problems with the templates. However, some users mentioned that there are some repetitions and some points could be summarized in fewer points. For instance, some users found it hard to distinguish between the System under Test in the Test Case description and the Specific Test System in the Test Specification description. Also, some users had difficulties in understanding the differences among the TC/TS/ES descriptions. The difference between the Test Criteria and the Target Metric was not described well. Regarding the templates, the users mentioned that they generally work fine for a Single-Test/Single-RI test case, but they become harder to use when multiple tests and/or RIs are introduced.

4.2 Results of Data Specification Questionnaire

All the users who participated in this questionnaire employed the Test Case Template, the Test Specification Template and the Experiment Specification Template of the HTD procedure, however one of the participants also employed the Qualification Strategy, the Experiment Realization Process (with RI-Database), the Design of Experiments Methodology and Statistical Analysis.

Concerning the "variable mapping", in case of differences between the user documentation and the realized experiment, no major difference was observed. However, some users had to modify the systems initially planned to adapt to the specification of the used equipment.

Regarding the "lab integration", it consisted of three steps; first, the interface of the new components with the lab, second, the real-time interaction among the labs, and, finally, the refinement of the lab set up based on previous experiments. For instance, one user had to deal with the interface of new components with the laboratory, while the previous project of that user was based on simulations. The problem was that those simulations were not directly exportable to the real laboratory, so they had to understand how the laboratory was structured and the control program was operated in order to obtain the correct results. They also had to change the initial measurement method to achieve greater precision in the data.

Speaking of how the result data was obtained from the experiment and stored by the users, some users stored the results in a Matlab file format in which case no data conversion was needed.

5 Advantages and Shortcomings of Holistic Validation Methodology

According to the questionnaire results, the majority of the RIs had not used any specific testing methodology before the definition of the HTD. Few partners refer to the Holistic Testing Procedure, developed also in ERIGrid, as a previous relative experience. Furthermore, it was mentioned that no clearly defined methodologies existed previously for Multi RI tests.

It was noticed that the HTD is relatively hard to use, especially for the first time, as it requires a thorough understanding of the various definitions and concepts used in the methodology. On the other hand, some users mentioned that certain templates that have been provided proved to be very useful, as there is also a guide that explains how to use them properly.

Based on the questionnaire feedback provided by the partners, the HTD methodology presents several benefits, which are evidently not only in comparisons to previous testing methodologies, but also in the implementations of single and multi-RI experiments.

5.1 Advantages of the Holistic Validation Methodology

While the majority of previously used testing methodologies to be compared to HTD consisted of custom methods and practices that were different for each individual RI, nevertheless HTD presents obvious strengths:

- It provides a complete overview of the test, and by following the testing procedure provided by ERIGrid (using the guidelines), the users can have heightened awareness of the context during the test setup.
- It helps to structure the planning of an experiment—providing the capacity to split complex test cases into individually achieved subtests and provides a means for the user to think about various critical aspects regarding the tests in a systematic

manner, easing the identification of the parameters that need to be accounted for when designing holistic tests.

- If performed properly, it can support documentation and communication, as the definition of basic concepts can contribute in improving multi-domain comprehension of complex validation activities and encourage interdisciplinary collaboration among teams.

Based on user's feedback, in experiments that include the participation of a single RI, employing the HTD methodology can provide key benefits to several parts of the testing process:

- By clarifying the Purpose of Investigations and Test and Experiment Specifications, it can lead to faster preparation of the TC, while the structured planning and execution of tests eases the communication process of test plans and results.
- The effort that the method requires for the definition and the organization of the tests proves to be quite helpful when there is a need to implement the same experiments and compare the results between two different RIs.
- In HIL/CHIL simulations, this method is mostly efficient when you have to develop a new component. What is more, the simulation and HIL experiments can extend the results of the pure hardware experiment.

Additionally, the utility of the HTD seems to be extended for tests that include the participation of more than one RI, in different physical locations. Let it be noted that any advantages present in single RI experiments also apply to the multi RI experiments, in addition to the following:

- It allows transparent translation between the system under test and the test setup and clearly determines the boundaries of the test setup among different partners. This can also assist in clarifying and distributing the workload between different RIs, avoiding misunderstandings during the experiment and improving interdisciplinary collaboration among teams.
- It leads to the standardization of test cases and the adoption of a common naming/signal exchange convention, improving the multi-domain understanding of complex validation activities.
- It offers the possibility to simultaneously use hardware assets at multiple RI without the need for transport of equipment or personnel, thus minimizing investment and transportation costs.
- Web services provided by a single RI (such as a state estimator) can be used by other RIs without any exchange of software.
- It can extend the validation capabilities of RIs and create advanced testing environments which cannot be implemented in a single RI mainly due to hardware limitations. It also extends the validation experience between partners (complementarity) and joint developments.

5.2   Shortcomings of the Holistic Validation Methodology

While constituting a noteworthy evolution of previous testing methodologies in several aspects, the HTD methodology does present certain insufficiencies, which the questionnaire aimed to identify. The answers of the participants converged to the following major points:

- The HTD methodology is still complex, long and the presence of several different terms and parameters can make it appear too "academic".
- Some of its concepts/terminologies in the templates can be difficult to understand and similarities in their definitions can lead to different interpretations and render the implementation of the methodology relatively cumbersome.
- It is missing a process for the collection and reporting of (sub) test results.
- It is in need of more support documentations and adjuvant tools, like a comprehensive user interface or pre-filled templates as examples.

6   Conclusion

This chapter presents a comprehensive evaluation of the system-level validation approach that was employed for the tests by the majority of the research teams and is known as the HTD method. Initially, a summary of the application of the method in two characteristic test case instances is presented, thus gauging its utility through examples and offering a first impression of its strengths and weaknesses. In order to perform a more thorough assessment of the method, based on the particular experiences of the HTD users, some questionnaires were designed and distributed, and the feedback was utilized to form a more structured evaluation.

Consequently, the results of two questionnaires are presented, one regarding the ERIGrid services and one concerning data specification. The former examines the quality of the services relevant to the holistic testing procedure, namely the holistic test case description, the system configuration description and the RIs database, while the latter searches opportunities for improvement in the fields of system configuration, definition and naming of variables and data exchange.

The chapter closes with the demonstration of the results of a questionnaire addressing the general advantages and shortcomings of the use of the HTD by individual RIs. The pros of the method in contrast to older methodologies are highlighted, and additional merits are described in experiments conducted both under single RI and multi RI status. The final section presents feedback related to the perceived imperfections and drawbacks of the method, which can evoke future improvements and can be used to determine goals for its subsequent development.

References

1. Heussen, K., Steinbrink, C., Abdulhadi, I.F., Nguyen, V.H., et al.: Erigrid holistic test description for validating cyber-physical energy systems. Energies **12**(14) (2019)
2. Uslar, M., Specht, M., Rohjans, S., Trefke, J., González, J.M.: The Common Information Model CIM: IEC 61968/61970 and 62325-A practical introduction to the CIM. Springer Science & Business Media (2012)
3. Lu, Y., Liu, D.: Key technologies of information exchange in electric utility based on IEC 61968. Przegląd Elektrotechniczny 88(11a), 276–282 (2012)

Education and Training Needs, Methods, and Tools

P. Kotsampopoulos, T. V. Jensen, D. Babazadeh, T. I. Strasser◉, E. Rikos,
V. H. Nguyen, Q. T. Tran, R. Bhandia, E. Guillo-Sansano, K. Heussen,
A. Narayan, T. L. Nguyen, G. M. Burt, and N. Hatziargyriou

1 Introduction

A need for new skills and expertise to foster the energy transition has risen, considering the increased complexity of Cyber-Physical Energy Systems (CPES). Tackling the contemporary significant challenges requires a skilled workforce and researchers with systemic/holistic thinking and problem-solving skills. At the same time, technological advances can revolutionise education by allowing the use of new technical tools.

P. Kotsampopoulos (✉) · N. Hatziargyriou
National Technical University of Athens, Athens, Greece
e-mail: kotsa@power.ece.ntua.gr

T. V. Jensen · K. Heussen
Technical University of Denmark, Roskilde, Denmark

D. Babazadeh · A. Narayan
OFFIS – Institute for Information Technology, Oldenburg, Germany

T. I. Strasser
AIT Austrian Institute for Technology, Vienna, Austria

E. Rikos
Centre for Renewable Energy Sources and Saving, Athens, Greece

V. H. Nguyen
Université Grenoble Alpes, INES, Le Bourget du Lac, France

CEA, LITEN, Le Bourget du Lac, France

Q. T. Tran
Université Grenoble Alpes, INES, CEA, LITEN, Le Bourget du Lac, France

R. Bhandia
Delft University of Technology, Delft, The Netherlands

E. Guillo-Sansano · G. M. Burt
University of Strathclyde, Glasgow, Scotland, UK

T. L. Nguyen
Université Grenoble Alpes, Grenoble INP, Grenoble, France

T. I. Strasser et al. (eds.), *European Guide to Power System Testing*,
https://doi.org/10.1007/978-3-030-42274-5_8

In this framework, educational and training needs addressing the higher complexity of intelligent energy systems are identified in this chapter. State-of-the art laboratory-based and simulation-based tools are employed to address these needs. Real-time hardware in the loop simulation for hands-on laboratory education is applied and its benefits are explained. The learners gain access to remote labs, that allow the remote monitoring and control of laboratory facilities. Simulation-based tools that focus on co-simulation support the systemic understanding, while interactive notebooks promote problem-solving skills. Webinars and training schools allow the use of the proposed methods and tools by larger audiences and the collection of feedback. The developed material and tools are publicly available[1] in order to promote the use and replicability of the approaches.

2 Learning Needs for Modern Power and Energy Education

Due to the increased complexity of intelligent power and energy systems, current and future engineers and researchers should have a broad understanding of topics of different domains, such as electric power, heat and definitely Information and Communication Technology (ICT) related topics. Appropriate education on modern topics is essential at university level, both for undergraduate and postgraduate studies, so that future engineers will be able to understand and tackle the challenges and propose/implement new methods. Recently, several universities have incorporated new courses in the undergraduate engineering curriculum or have enriched their existing courses with more modern material, while some universities have created dedicated master courses with relevant topics. The instruction is performed using traditional methods, such as class lectures, but also with programming, advanced simulations [25] and laboratory exercises [5, 10], that occasionally include the application of advanced learning methods such as problem-based learning and experiential learning [6, 14].

Moreover, the ongoing training of current professional engineers on modern topics is important. In some cases, professionals may tend to be hesitant of change and prefer to use proven technologies and methods. By taking part in proper training, professional engineers can better understand the benefits of modern solutions and ways to apply them in order to improve their daily work. Specifically, power system professionals frequently lack thorough understanding of ICT topics, while ICT professionals often find it hard to understand the operation of the power system. As these areas are closely connected due to the emergence of intelligent power and energy systems, it is important to create links between them. Obviously, a thorough understanding of all domains (electric power, heat, ICT, automation, etc.) is difficult to achieve, however an understanding of the fundamentals of each area, without sacrificing the expert focus in each particular field, will become increasingly important.

---

[1] https://erigrid.eu/education-training/.

The same applies to researchers who are working to find solutions beyond the state of the art.

The future experts require the relevant insight and abilities to design and validate solutions for CPES. These abilities are facilitated by both generic engineering and cross-disciplinary technical competences. The generic competences include the conception, design, implementation and operation of systems (e.g. based on the Conceive Design Implement Operate (CDIO) skills catalogue [4]), which can be supported, for example, by project-oriented teaching methods. With increasing problem complexity, systems-oriented skills need to become strengthened, such as problem decomposition, abstraction and multi-disciplinary coordination of engineering challenges.

Cross-disciplinary learning is also required as the integration and interdependency of software and hardware systems is increased. Engineering students who aim to design and work with CPES solutions like complex control, data analytics, supervisory and decision support systems, require an increased level of programming and system design competences, as well as a pragmatic view on the applicability of methods. This means that some familiarity with domain specific system architectures and description methods is useful (e.g. reference architectures such as Smart Grid Architecture Model (SGAM) [8, 19]). Moreover, basic familiarity with distributed software system problems is important, as they are not addressed sufficiently within contemporary engineering education. In addition, simulation-based tools are useful for the emulation and understanding of physical behaviour and cyber-physical system couplings. More information on contemporary learning needs can be found in [12].

Summarizing, the following learning needs in the domain of intelligent power and energy systems are identified:

- Understanding the physical layer of CPES (especially topics related to distributed energy resources), including the interconnected sub-systems and components.
- Understanding automation and control systems.
- Understanding communication networks.
- Understanding optimization, data analytics and artificial intelligence.
- Understanding the mutual interactions/influences amongst components and domains.

Therefore, a holistic understanding of the physical and the cyber part of intelligent power and energy systems is necessary in order to design and develop a future reliable and sustainable energy system. This should be reflected in current and future education and training. In this direction, laboratory and simulation-based tools and methods are presented in the following sections to advance education and training in the smart grid era.

3 Laboratory Education

Laboratory education provides a link between theory and real world offering valuable practical experience to students. In this section, new trends in laboratory education are presented, such as the use of real-time Hardware in the Loop (HIL) simulation and remote laboratories, including representative examples.

3.1 Real-Time Simulation for Laboratory Education

Laboratory education on power systems is usually performed with simulation software [7] or less frequently with dedicated hardware setups [15, 16, 20, 22]. On the other hand, laboratory education on power electronics and electric machines is typically performed with hands-on exercises using physical models or real hardware, as the focus is on the component level. The limited use of real hardware on power system education is obviously due to the difficulty and cost of having a realistic power system setup in the lab (including generators, transformers etc). As a result, small educational hardware setups usually perform specific functions and cannot be easily used for a wide range of experiments.

Real-time HIL simulation merges simulation and hardware testing providing hardware experience to the students, while exploiting the advantages of digital simulation. The following features of HIL simulation are beneficial for educational purposes:

- The students face a real-time system (like a SCADA), where they can perform actions and monitor the operation in realistic conditions. The flexibility, ease of modelling and designing test scenarios of digital simulation are maintained.
- The connection of real hardware devices such as inverters of Distributed Generation (DG), microgrids, relays can be realised, so that students can observe the operation of real apparatus. Measurement of actual magnitudes and control of real devices is a valuable experience.
- Components that are not available in the lab (e.g. transformer, diesel generator) can be simulated in real-time and their interaction with hardware devices can be studied.
- Challenging tests, such as faults, can be performed safely and conveniently in a real-time simulation environment without hazardous effects or equipment stress. The type, duration and location of faults can be easily modified by the students, which would be difficult in a real hardware setup.

Controller Hardware-in-the-Loop (CHIL) simulation has been used several times for laboratory education [3, 24], however the potential of Power Harware-in-the-Loop (PHIL) experiments could be perfectly explored during the ERIGrid project since it required a new and more complex technical approach.

3.1.1   Laboratory Exercise Examples

The laboratory exercises aim to introduce university students to the world of real-time simulation and its various applications in the power system domain. Several modules have been created which start from an introduction to real-time simulation, leading to more advanced modelling and interfacing techniques, using a hands-on approach. At first, the students become familiar with real-time simulation software (RSCAD of RTDS) by executing several examples. Interfacing with other software tools is introduced (e.g. Matlab and Python), highlighting the co-simulation possibilities. The possibility of developing custom component models and the small time-step modelling of power electronic devices are explained. Next, the use of communication protocols (e.g. IEC 61850) is highlighted, along with interfacing techniques with hardware equipment. Finally, practical examples of modelling of Photovoltaic (PV) systems and High Voltage DC (HVDC) links are provided.

A laboratory exercise for explaining voltage control of distribution networks with distributed generation is discussed in detail, where the students have the opportunity to monitor and control actual equipment. An overview of the PHIL/CHIL setup of the experiment is shown in Fig. 1. A hardware PV inverter and a load bank are connected to a simulated weak distribution network fed by a transformer equipped with an On-Load Tap Changer (OLTC). The students control the active power of the PV simulator by changing the irradiation for a given I-V curve via its software environment. While keeping the load low, they steadily increase the active power of the PVs from zero to nominal and observe the voltage rise occurring at the inverter's terminal. The students try to solve this overvoltage problem and suggest as a solution the reactive power absorption by the PV inverter. They send reactive power absorption set-points to the PV inverter via its software interface, monitor the voltage and validate its effect. In this way, the need for DG to support the grid by providing ancillary services is highlighted. More complex solutions are gradually demonstrated and explained, such as the application of coordinated voltage control that requires the existence of a telecommunication network, the solving of an optimization problem etc.

The laboratory exercises are designed according to the principles of experiential learning, based on Kolb's four-stage learning cycle [11]. According to that cycle, the initial concrete experience is being elaborated and reflected upon (reflective observation) to enable the learner to reach an abstract conceptualization, which is the third stage of the learning cycle. That abstract concept is being applied in real life situations (active experimentation) so that a new concrete experience emerges which is elaborated and reflected upon and so on. Accordingly, during the classroom lectures the students are taught fundamentals of power system operation. During the experiments, the DG integration topics are offered to the students directly in the lab (concrete experience) without in-depth theoretical knowledge. After each experiment, suitable questions on real problems are posed to facilitate understanding, taking into account the student's existing knowledge (reflective observation). Guided conversations with the students or direct instruction, when considered necessary, lead to new concepts (abstract conceptualization). On this ground, new experiments are performed (active experimentation and new concrete experience). The reports at the

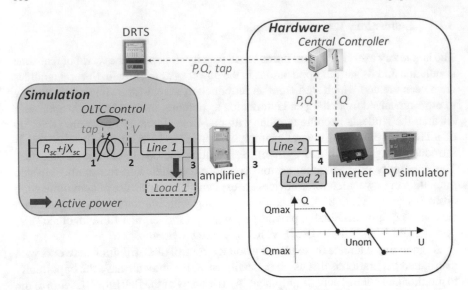

Fig. 1 Voltage control by DG, OLTC and centralised coordinated control (PHIL and CHIL simulation) [14]

end of the session aim to reflect the laboratory experience (reflective observation). More information on the laboratory exercises can be found in [14].

3.2 Remote Laboratories

Remote labs are gaining significant attention for educational purposes, as they allow the user to connect remotely to actual laboratory infrastructure, obtain measurements and control devices [1, 9]. Two remote laboratory applications are presented next.

3.2.1 Voltage Control

The remote lab for voltage control provides online access to actual laboratory equipment, allowing measurement and control via the laboratory SCADA. The laboratory setup includes a PV inverter that is connected to the utility grid via a long low voltage line. The user can control the active and reactive power of the inverter and monitor the resulting voltage at the actual hardware setup, from the web-based Graphical User Interface (GUI), shown in Fig. 2.

Moreover, a virtual lab has been developed that uses a mathematical representation of the system and is also available as a web-based tool. The remote lab is more realistic than the virtual lab, as the operation of the real system is observed, providing a more meaningful experience to the user. In addition, phenomena such as noise, equipment

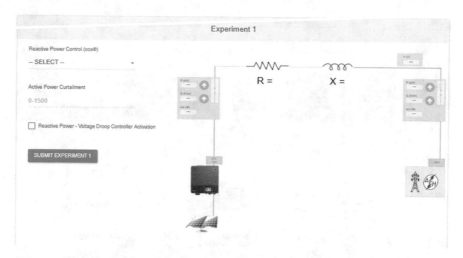

Fig. 2 Graphical user interface of the remote lab for the voltage control experiment

inaccuracies, communication delays, etc which are inherent in a real-life system are taken into account at the remote lab. On the other hand, the use of the remote lab presents some challenges. Most importantly, only one user can typically have access to the remote lab application at a time (because there is only one laboratory setup available to control), whereas the virtual lab can be used by simultaneously by a large number of users at the same time through the online platform. Therefore, it is more difficult to offer the remote lab to a wide audience. Moreover, for safety reasons it is recommended that laboratory staff monitors the process of the experiment and communicates with the user if necessary.

3.2.2 Microgrid Balancing

In order to facilitate the understanding of concepts such as multi-domain experiments, interoperability of control devices and real-time simulation of physical components, the Microgrid Balancing remote lab application was designed and implemented. The basic idea of this experiment is that a user can remotely connect to the experimental microgrid and interact with the SCADA system in order to achieve a power balancing operation based on specific market policy. In the setup the battery, load, and grid connection are all physical components, whereas the PV unit is simulated (in MATLAB/Simulink) in order to introduce the analytical mathematical models of the PV system (e.g. calculation of I–V characteristic, Maximum Power Point, injected AC power). The input signals to the PV model are global horizontal irradiance and ambient temperature from real-time measurements. Figure 3 provides an overview of the Simulink model blocks.

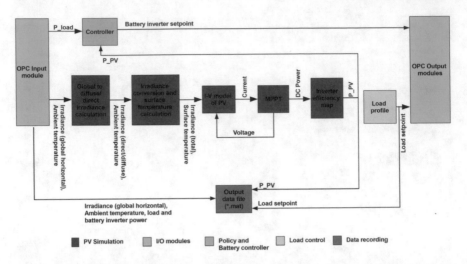

Fig. 3 Analytical illustration of the control blocks of the remote lab for microgrid balancing

The controller makes use of a simple strategy which chooses a power set-point for the batteries according to the power values of the PV and loads. The control is divided into two scenarios, named as "sell" and "buy priority". In "sell priority" the injection of the PV power surplus to the grid is prioritised, while in "buy priority" the battery covers the power imbalance by absorbing any PV power surplus.

4 Simulation-Based Tools

Given the cross-disciplinary nature of intelligent power and energy systems as outlined above, students should be exposed to a wide set of tools and concepts related to different domains. Thus, new educational methods and tools must be developed, capable of bringing the knowledge of different domains together and allowing the students to understand the coupling and interaction of elements within intelligent solutions.

It is clear that simulations will play an important role in the design, analysis and testing process of new solutions. It is therefore natural that students should learn to use domain-specific simulation tools, both in standalone and in co-simulation setups. The students should also understand the limitations of such tools. Moreover, methods that support students to bridge the gap of theory and application are required.

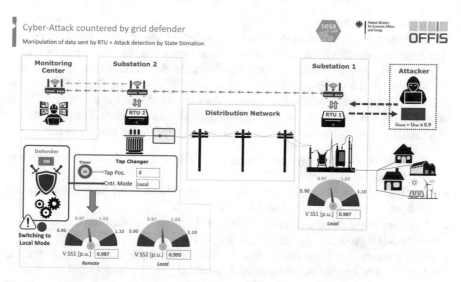

Fig. 4 Graphical user interface of the cyber-resilience tool

4.1 Co-simulation Tools

4.1.1 Mosaik-Based Co-simulation

The goal of mosaik co-simulation framework is to provide researchers with an easy-to-use yet powerful tool for simulation-based testing [21, 23]. It provides a flat learning curve for learners/researchers wishing to test their simulations in common scenarios with others, allowing intuitive co-modelling of various power system scenarios. Next to its research applications, mosaik is used for training and education in the field of CPES. In a practical course, students learn to plan, execute, and analyse co-simulation-based experiments. The target audience includes researchers and students from domains such as computer science, environmental modelling and renewable energy. They learn how to develop models of electrical components and integrate them into smart grid scenarios. They also learn to develop distributed agent-based control algorithms for smart grids. These developments can be done in individual modelling environments and are co-simulated using mosaik. The goal is to analyze the requirements for real-time performance, accuracy, resource utilization and the reliability of the simulation results.

4.1.2 Cyber-Resilience Tool

The increased penetration of active components and digitalization intensifies the system complexity, resulting in higher risk of ICT incidents [18], thereby expanding the scope of cyber threats. The Cyber-Resilience Tool is an educational tool to demonstrate how cyber vulnerabilities could affect an electrical distribution grid. It also shows a possible defensive action against cyber-attacks. The tool has a GUI (shown in Fig. 4) with which the users can perform certain attacks and investigate their impact on the system; both with and without defensive measures. Since smart grids consists of multiple domains, one of the main challenges is the integration of different tools i.e. co-simulation, so as to analyse different domains. The tool consists of a real-time simulation environment including the following components:

- Modelling of the power system feeder, e.g. loads, lines, busbars and transformer (*ePHASORSIM* from OPAL-RT).
- Modelling of the tap changer controller of the transformer and the defensive mechanism of the system (*eMEGASIM* from OPAL-RT).
- Modelling of the communication infrastructure between the substations using IEC 60870-5-104 protocol (*EXata* from Scalable Network Technologies).
- Communication protocol translation and modelling of payload alteration options, i.e. cyber-attacker (*Virtual Remote Terminal Unit (vRTU)* from OFFIS).

4.1.3   FMU-as-a-Service Approach

The Functional Mock-up Interface (FMI) is a standard in co-simulation that allows interoperability among models from different domains. The Functional Mock-up Unit (FMU—the basic brick of FMI standard) encloses the dynamic model and generates compiled C code, which can be integrated into other environments as a black box. The deployment of FMUs in a co-simulation framework is however problematic for novice users without extensive informatics background. This limits the development of FMU and hinders the learning curve as well as the utilization of FMI by new users.

In this context, a software tool has been developed allowing the delivery of FMU in a Software-as-a-service manner, named as FMU-as-a-service. The server is developed on Django with PyFMI as the solver. Results are available in JSON-CSV-HTML or can be represented at the graphical web interface. With the proposed tool, no installation is required from the user side, and an FMU could be executed without further requirements (e.g. toolbox, solver request, etc.). Moreover, the platform allows multiple users working on the same model at the same time. The user can also deposit and encrypt (RSA) their FMU model to the server to keep the code confidential.

One of the main objectives of this software is to help students and interested users to understand the functionality and structure of a functional mock-up unit and to become familiar with planning, executing and evaluating simulation-based experiments. Initially developed to serve only in the smart building validation domain, it can be used for various courses involving (co)simulation, such as mechatronics or robotics or eventually complex cyber-physical systems. Moreover, the software can be used to provide a comprehensive course on different type of simulations, methods of computation and on their interaction.

4.2   Interactive (Jupyter) Notebooks

A main issue with smart grid validation, especially for large-scale systems, is the complexity of the resulting simulation. When instructing in these complex simulations, a large portion of time is spent trying to couple the overall learning of the activity into the code that students must write or engage with. As the complexity of the code base inevitably increases, students who are not used to dealing with this aspect may feel lost in the aspects of the code itself. In the worst case, this may cause the student to miss the high-level learning involved. One way to address this issue is to apply interactive notebooks in instruction.

Jupyter notebooks are a merge between a standard text book and what real programming in python looks like [2]. The notebooks can be built from explanatory text and figures, while a full Python kernel allows the student to execute python code. The use of these notebooks provides a way to narrow the gap between theoretical concepts and application by setting up code examples, where the student is able to directly see examples of the theoretical concepts explained in the text. Notebooks of this kind can be developed to cover a wide spectrum of intelligent energy systems concepts.

By constructing a framework where students can focus on a problem to solve, instead of dealing with issues related to programming, they are able to better absorb and understand the core course concepts. The learning objectives of the developed notebooks are as follows:

- Design a testing procedure to validate a "black box" algorithm.
- Recognize the importance of statistical Design of Experiments to qualifying tests.
- Apply Design of Experiments to evaluate the performance of a "black box" algorithm.

As an example of the application of this tool, a pair of notebooks were given to the students for hands-on experience with a co-simulation environment and Design of Experiments [17]. The example used consisted of a typical Home Energy Management System in which a scenario is built up step by step, to include a house with solar panels, battery and a controller that can connect the house to the grid or discharge the battery. All the components and interconnections were modelled in behind-the-scenes python scripts and the simulation ran using the python module mosaik. An example of the final notebook can be seen in Fig. 5.

5   Outreach Activities

Outreach educational/training activities, such as the delivery of webinars, training schools and workshops promote the use of the proposed methods and tools by larger audiences. Insights from the delivery of several webinars, training schools and workshops are provided below.

5.1 Webinars

The use of webinars (i.e. web seminars) as an e-learning environment is receiving more and more attention [26, 27]. The possibility of addressing large audiences, conducting live exercises/experiments to engage with the audience and also interacting with the presenters through questions and answers, render webinars a valuable educational tool. Moreover, the activity presented at the webinar is typically recorded and distributed to the participants and also made publicly available through video sharing platforms for later viewings at the learner's own pace. In this way a larger audience than the initial participants are able to benefit from the webinar.

In the contemporary cyber-physical environment, when the learning topics are related to the introduction of complex methods and tools (e.g. advanced testing and simulation), the webinar is an efficient option to achieve the learning objectives. For example, cases in which a webinar was found to be an effective learning environment are:

1. Introduction to a new software tool, where the participants have no previous experience.
2. Performance of live demonstrations of simulations and also laboratory tests.
3. Understanding co-simulation possibilities for cyber-physical systems.

It should be noted that feedback from the webinar participants can be easily obtained and analysed in order to improve the learning process.

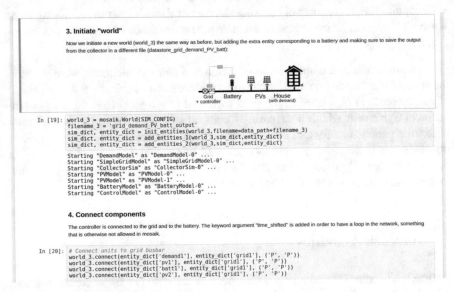

Fig. 5 Example of a Jupyter notebook, where markup language and Python code can be used to showcase complex examples

5.2   *Training Schools and Workshops*

Training schools, such as summer schools or winter schools, typically target PhD candidates, early stage researchers, and young professionals and have a duration that allows the in-depth analysis of the given topic. On the other hand, a workshop has a shorter duration and is typically focused on giving necessary background for application. While a training school or workshop is necessarily limited in the scope of the audience that can benefit from it, a main advantage is that the participants can have close interaction with the instructors. Thus, training schools and workshops are particularly relevant when the learning goal is to:

- Familiarise with the use of a local laboratory infrastructure.
- Apply advanced theoretical methods, especially when few standard learning resources exist for the smart grid context.
- Familiarize with complex co-simulation methods.
- Design and evaluate systems for which the failure modes are too numerous to anticipate.

As laboratory-based methods play an important role in the analysis of CPES, training schools and workshops proved to be efficient educational/training settings. Hands-on approaches were applied which included individual work (e.g. familiarization with a specific software) combined with team work (e.g. joint modelling, fine tuning and running simple experiments). Live demonstrations are beneficial when an experiment is too complex for a hands-on activity and can better demonstrate the capabilities and inspire the participants. A well-balanced set of lectures, hands-on laboratory work and visits to industrial installations was found to be an ideal combination for a successful event, as long as the activities integrate in a structured way into the event's overall purpose.

6   Conclusions

The emergence of intelligent solutions in the domain of power and energy systems opens new possibilities but poses new challenges, rendering appropriate education and training methods for students and engineers increasingly important. A broad understanding of several domains is necessary to deal with the increased complexity and diversity, including electric power systems, automation, ICT, thermal systems etc. This chapter identifies upcoming educational needs and requirements in this rising complex environment. It is explained that the validation of complex systems is a multi-stage process, while systems-oriented skills and cross-disciplinary learning needs to be cultivated. In order to cover the distance between theory and hands-on practice, coding and laboratory education is beneficial. As the required field of knowledge is too broad, educational methods such as experiential learning and problem-based learning can prove to complement effectively the traditional teaching methods.

In this framework, state of the art laboratory-based and simulation based-tools and methods were presented, accompanied by representative examples. The benefits of real-time simulation for educational purposes were clearly explained, highlighting the provision of hands-on experience to students. The remote labs give users the possibility to experience lab conditions by gaining online access to actual laboratory installations providing a realistic experience. Simulation-based tools are efficient ways to educate on CPES topics such as co-simulation. They can help bridging the gap between theory and application, can be used in different settings (e.g. classroom sessions, e-learning, demonstrations at workshops) and support blended learning. The delivery of webinars and the organization of training schools proved to be an effective way to educate students, researchers and professionals on emerging topics. It is shown that a holistic approach using advanced tools and methods can advance education and training in the field of CPES. More information about the developed tools and methods can be found in [13].

References

1. Aydogmus, Z., Aydogmus, O.: A web-based remote access laboratory using scada. IEEE Trans. Educ. **52**(1), 126–132 (2008)
2. Cardoso, A., Leitão, J., Teixeira, C.: Using the Jupyter notebook as a tool to support the teaching and learning processes in engineering courses. In: International Conference on Interactive Collaborative Learning, pp. 227–236. Springer (2018)
3. Celeita, D., Hernandez, M., Ramos, G., Penafiel, N., Rangel, M., Bernal, J.D.: Implementation of an educational real-time platform for relaying automation on smart grids. Electr. Power Syst. Res. **130**, 156–166 (2016)
4. Crawley, E.F., Malmqvist, J., Lucas, W.A., Brodeur, D.R.: The CDIO syllabus v2. 0. an updated statement of goals for engineering education. In: Proceedings of 7th International CDIO Conference, Copenhagen, Denmark (2011)
5. Deese, A.S.: Development of smart electric power system (seps) laboratory for advanced research and undergraduate education. IEEE Trans. Power Syst. **30**(3), 1279–1287 (2014)
6. Felder, R.M., Brent, R.: The abc's of engineering education: Abet, bloom's taxonomy, cooperative learning, and so on. In: Proceedings of the 2004 American Society for Engineering Education Annual Conference and Exposition, vol. 1 (2004)
7. Georgilakis, P.S., Orfanos, G.A., Hatziargyriou, N.D.: Computer-assisted interactive learning for teaching transmission pricing methodologies. IEEE Trans. Power Syst. **29**(4), 1972–1980 (2014)
8. Gottschalk, M., Uslar, M., Delfs, C.: The Use Case and Smart Grid Architecture Model Approach: The IEC 62559-2 Use Case Template and the SGAM Applied in Various Domains. Springer, Berlin (2017)
9. Gustavsson, I.: Remote laboratory experiments in electrical engineering education. In: Proceedings of the Fourth IEEE International Caracas Conference on Devices, Circuits and Systems (Cat. No. 02TH8611), pp. I025–I025 (2002)
10. Hu, Q., Li, F., Chen, C.F.: A smart home test bed for undergraduate education to bridge the curriculum gap from traditional power systems to modernized smart grids. IEEE Trans. Educ. **58**(1), 32–38 (2014)

11. Kolb, D.A.: Experiential Learning: Experience as the Source of Learning and Development. FT Press (2014)
12. Kotsampopoulos, P., Hatziargyriou, N., Strasser, T.I., Moyo, C., et al.: Validating intelligent power and energy systems - a discussion of educational needs. In: Marik, V., Wahlster, W., Strasser, T., Kadera, P. (eds.) Industrial Applications of Holonic and Multi-Agent Systems. HoloMAS 2017. Lecture Notes in Computer Science, vol 10444. Springer, Cham (2017)
13. Kotsampopoulos, P., Maniatopoulos, M., Tekelis, G., Kouveliotis-Lysikatos, I., et al.: D-NA4.2a Training/education material and organization of webinars. Deliverable D4.3, ERIGrid Consortium (2018)
14. Kotsampopoulos, P.C., Kleftakis, V.A., Hatziargyriou, N.D.: Laboratory education of modern power systems using phil simulation. IEEE Trans. Power Syst. **32**(5), 3992–4001 (2016)
15. Kuzle, I., Havelka, J., Pandžić, H., Capuder, T.: Hands-on laboratory course for future power system experts. IEEE Trans. Power Syst. **29**(4), 1963–1971 (2014)
16. Maza-Ortega, J.M., Barragán-Villarejo, M., de Paula García-López, F., Jiménez, J., Mauricio, J.M., Alvarado-Barrios, L., Gómez-Expósito, A.: A multi-platform lab for teaching and research in active distribution networks. IEEE Trans. Power Syst. **32**(6), 4861–4870 (2017)
17. van der Meer, A.A., Steinbrink, C., Heussen, K., Morales Bondy, D.E., et al.: Design of experiments aided holistic testing of cyber-physical energy systems. In: 2018 Workshop on Modeling and Simulation of Cyber-Physical Energy Systems (MSCPES), pp. 1–7. IEEE (2018)
18. Narayan, A., Klaes, M., Babazadeh, D., Lehnhoff, S., Rehtanz, C.: First approach for a multi-dimensional state classification for ICT-reliant energy systems. In: International ETG-Congress 2019; ETG Symposium, pp. 1–6. VDE (2019)
19. Neureiter, C., Engel, D., Trefke, J., Santodomingo, R., Rohjans, S., Uslar, M.: Towards consistent smart grid architecture tool support: from use cases to visualization. In: IEEE PES Innovative Smart Grid Technologies, Europe, pp. 1–6 (2014)
20. Rasheduzzaman, M., Chowdhury, B.H., Bhaskara, S.: Converting an old machines lab into a functioning power network with a microgrid for education. IEEE Trans. Power Syst. **29**(4), 1952–1962 (2014)
21. Rohjans, S., Lehnhoff, S., Schütte, S., Scherfke, S., Hussain, S.: Mosaik-a modular platform for the evaluation of agent-based smart grid control. In: IEEE PES ISGT Europe 2013, pp. 1–5. IEEE (2013)
22. Santoso, S., Lwin, M., Ramos, J., Singh, M., Muljadi, E., Jonkman, J.: Designing and integrating wind power laboratory experiments in power and energy systems courses. IEEE Trans. Power Syst. **29**(4), 1944–1951 (2014)
23. Schütte, S., Scherfke, S., Tröschel, M.: Mosaik: A framework for modular simulation of active components in smart grids. In: 2011 IEEE First International Workshop on Smart Grid Modeling and Simulation (SGMS), pp. 55–60. IEEE (2011)
24. Srivastava, A., Schulz, N.: Applications of real time digital simulator in power system education and research. In: American Society for Engineering Education. American Society for Engineering Education (2009)
25. Strasser, T., Stifter, M., Andrén, F., Palensky, P.: Co-simulation training platform for smart grids. IEEE Trans. Power Syst. **29**(4), 1989–1997 (2014)
26. Verma, A., Singh, A.: Leveraging webinar for student learning. In: 2009 International Workshop on Technology for Education, pp. 86–90 (2009)
27. Wang, S.K., Hsu, H.Y.: Use of the webinar tool (elluminate) to support training: The effects of webinar-learning implementation from student-trainers' perspective. J. Interact. Online Learn. **7**(3), 175–194 (2008)

Summary and Outlook

T. I. Strasser⊙**, E. C. W. de Jong, and M. Sosnina**

1 Conclusions

The expected large-scale roll out of Cyber-Physical Energy System (CPES) products and solutions during the next few years requires a multi-disciplinary understanding of several domains. The validation of such complex solutions gets more important as in the past and there is a clear shift from component-level to system-level testing. An integrated, cyber-physical systems-based, multi-domain approach for a holistic testing of smart grid solutions is currently still missing which is addressed by the ERIGrid approach [3].

Four main research priorities have been identified in this pan-European project to tackle the shortcomings in today's validation and testing of power systems and corresponding components. The research focus is put onto the development of a holistic validation methodology and the improvement of simulation-based methods, Hardware-in-the-Loop (HIL) approaches, and lab-based testing, which can be combined in a flexible manner. The integration and online connection of power systems/smart grid laboratories is also a challenging research and development task in ERIGrid. All these activities need to be supported by the training of researchers and power system professionals [3].

With the in the book described integrated pan-European Research Infrastructure (RI) approach in ERIGrid the following improved methods, services, and tools are being made available by the consortium members [3]:

T. I. Strasser (✉)
AIT Austrian Institute of Technology, Vienna, Austria
e-mail: thomas.strasser@ait.ac.at

E. C. W. de Jong
KEMA B.V., Arnheim, The Netherlands

M. Sosnina
European Distributed Energy Resources Laboratories (DERlab) e.V., Kassel, Germany

© The Author(s) 2020
T. I. Strasser et al. (eds.), *European Guide to Power System Testing*,
https://doi.org/10.1007/978-3-030-42274-5_9

- Structured approach for defining a holistic description of validation and testing needs together with corresponding experiment descriptions,
- Improved co-simulation based approach with corresponding Functional Mock-up Interface (FMI)-based model libraries addressing the multi-domain and cyber-physical character of smart grid configurations,
- Improved HIL-based concepts analysing system-integration aspects of smart grid components,
- Possibility to couple co-simulation with lab experiments addressing system-integration aspects,
- Evaluation of different Information and Communication Technology (ICT) and automation architectures, control concepts,
- Support for rapid-prototyping of components and analysing their behaviour in power systems,
- Support for evaluating different smart grid system configurations, and
- A set of educational material supporting the training of power systems professionals.

Summarizing, engineering and validation support will be critical for successful development of future CPES applications and solutions. Without the proper tool support many of the tasks will require immense manual efforts and will require engineers educated in multiple domains (energy system physics, ICT, automation and control, cyber-security, etc.). The available ERIGrid results provide a step forwards in the right direction, but more research and development efforts are still needed in the years to come as outlined below.

2 Future Work

In fact, current research show that already many aspects needed for a better engineering and validation of CPES are available. Nevertheless, many issues are still open and since the advancement of CPES technologies is still ongoing, new needs are constantly appearing. Especially also the integration of the electric energy system with other domains (thermal, gas, water/waste water, transportation, etc.) into a smart energy system [1] requires additional efforts. Therefore, the following list is an attempt to point out possible research and development directions that still needs to be explored from CPES validation point of view [2, 4]:

- Advanced RIs need to be developed which focus on the integration of different power and energy systems related areas (market issues, thermal topics, electric vehicle, etc.),
- A simplified access and corresponding services (facilitate future access by remote operation and coupling of both virtual and physical RI, etc.) to smart grid, smart energy systems, and renewable related RIs addressing challenging user needs is required,

- Domain-specific adaptations of previously developed abstract validation procedures and corresponding concepts, methods, and tools are required to address advanced applications (low-inertia grids, microgrids, hybrid grids, etc.),
- Common and well understood reference scenarios, use cases, and test case profiles for smart energy systems need to be provided to power and energy systems engineers and researchers; also, proper validation benchmark criteria and key performance indicators as well as interoperability measures for validating smart grids and smart energy systems need to be developed, extended, and publically shared with domain professionals,
- A standardization of multi-domain CPES-based evaluation and testing procedures is necessary,
- Professionals, engineers, and researchers understanding smart grid and smart energy systems configurations in a multi-domain and cyber-physical manner addressing the upcoming energy transition need to be educated and trained on a broad scale.

The above listed open research and development issues are tackled by the successor project ERIGrid 2.0[1] which will be executed during the next years where several results are being provided open access to the domain of power and energy systems.

References

1. Mathiesen, B.V., Lund, H., Connolly, D., Wenzel, H., Østergaard, P.A., Möller, B., Nielsen, S., Ridjan, I., Karnøe, P., Sperling, K., et al.: Smart energy systems for coherent 100% renewable energy and transport solutions. Appl. Energy **145**, 139–154 (2015)
2. Neis, P., Wehrmeister, M.A., Mendes, M.F.: Model driven software engineering of power systems applications: literature review and trends. IEEE Access **7**, 177761–177773 (2019)
3. Strasser, T., Pröstl Andrén, F., Widl, E., Lauss, G., et al.: An integrated pan-European research infrastructure for validating smart grid systems. e & i Elektrotechnik und Informationstechnik **135**(8), 616–622 (2018)
4. Strasser, T.I., Andrén Pröstl, F.: Engineering and validating cyber-physical energy systems: needs, status quo, and research trends. In: International Conference on Industrial Applications of Holonic and Multi-Agent Systems, pp. 13–26. Springer (2019)

---

[1] https://erigrid2.eu.